High Tech Trademarks

John Mendenhall

John Mendenhall

Art Direction Book Company

ISBN: 0-88108-0241
LCC #85-061876

Printed in the United States of America

Published by
Art Direction Book Company
10 East 39th Street
New York, NY 10016

To William Lee

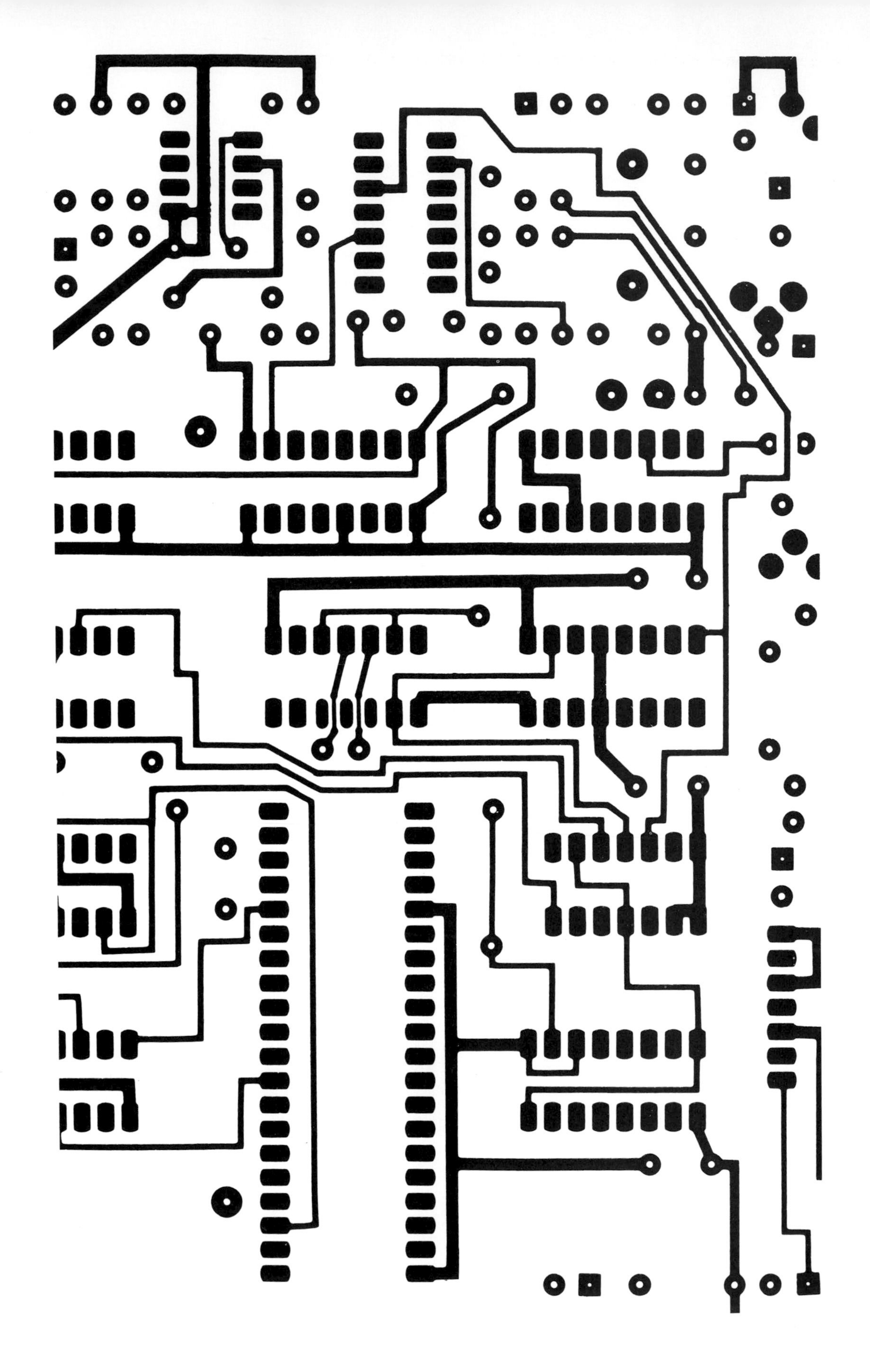

Introduction

"Everything seems to beep these days."
— Cynthia Smith

Even trademarks. The logos in this collection represent many of the most influential corporations involved in the high technologies of electronics, semiconductors, lasers, computer hardware and software. Gathered over a period of eighteen months, this book is the first effort to assemble the symbols of leading edge industries into one source.

Over 2500 requests were made directly to the companies as well as design offices which either specialize in, or do a significant amount of work for high technology clients. The resulting compendium is one of the most substantive records of the symbols of a new technological era.

As society becomes increasingly automated and computerized, the signs and symbols which make up our visual landscape change as well. The tools of high technology have altered how we transmit and perceive information. The handcrafted image, replicated by the printing press, is gradually being supplanted by new methods of graphic communication. Innovations such as computer-assisted design, video display terminals, jet spray and laser printing are expanding the ways by which designers conceive and create the artifacts of our culture.

In ten years will drafting instruments still exist, or might they be only available in antique shops, alongside the slide rules and adding machines of the 1970s? To students of design it may seem surprising that at one time felt markers did not exist: designers conceptualized ideas with pastels, smeared and blended with their fingers. Messy, but they did the job.

The invention of the felt tip marker in the 1950s changed all that, and accordingly designers changed the way they designed. Angular, hard-edged graphic design made by a hard medium on paper gave way to a smoother, more fluid image made by the liquid medium of the marker. The vibrant colors of 1950s graphics were as much a result of the upbeat mood of the times as the result of the brilliantly saturated colors available in the designer's marker set.

The move from pastels to markers as a conceptualizing medium was insignificant compared to the quantum leap made possible by the high technology of today. This changeover has just started to take place, thus its impact can only be speculative. By necessity, design methodology shifts as technology advances, and new movements in design develop as often in concurrence with this advance as in reaction against it. The machine aesthetic of Art Deco was a reaction against Art Nouveau's overindulgence with nature; the intuitiveness of Post Modern graphic design is a strong reaction against Swiss formality. It will be interesting to observe how designers will react to a technology which makes it possible to automate *all aspects* of their work, from concept to realization, subjugating eye-hand skills for eye-CRT dexterity.

Some designers will always prefer the tools of the "pre-technology" era: cold type with its precise form, for instance, is still used by those who find the more common digitized typography imperfect. Convincing designers to give up their layout pads for computer paint systems may be a more difficult task than earlier envisioned. Despite of (or perhaps as a *result of)* advances in graphic art technology, it is imperative that the human element always remain in the design process. As Milton Glaser so aptly put it: "Computers are to design what microwave ovens are to cooking."

Today's technology derives its strength from its smallest component: the semiconductor chip. A silicon wafer into which is etched layers of circuitry, the chip is the heart and brains of today's electronics. The integrated circuit (IC), whether magnified from a semiconductor or rendered in its schematic form, is one of the great icons of modern technology. The IC is similar in appearance to a geographic map, the many lines comprising a complex system capable of processing information in the form of electrical impulses. How they operate

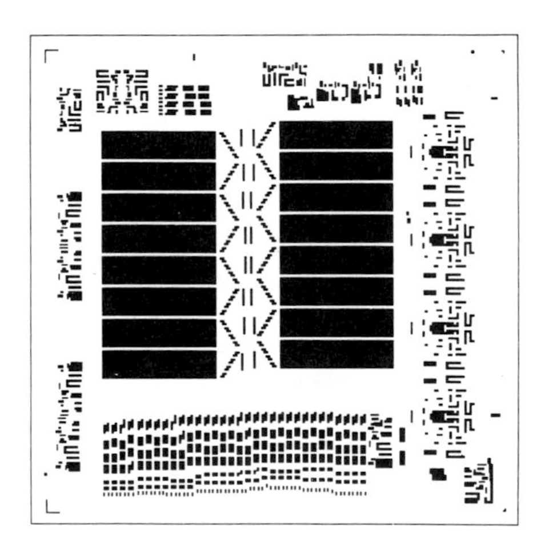

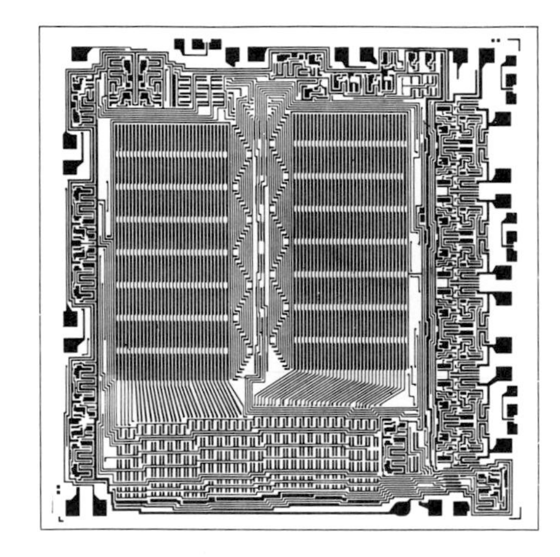

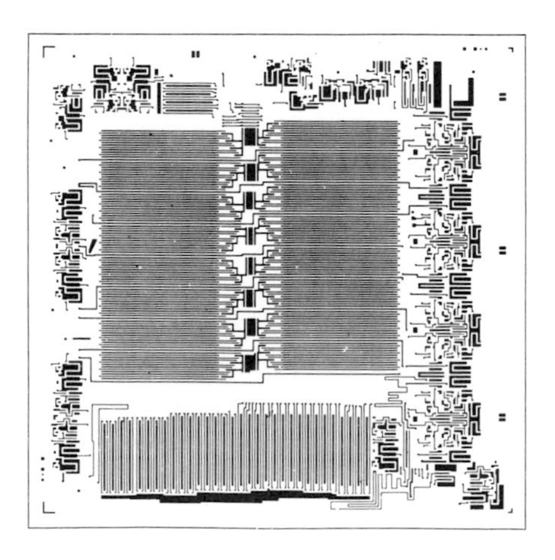

Photographic masks used in the etching of a semiconductor memory chip (ROM)

and what exactly they do is a mystery to most of us, yet their impact on society for what they make possible has created a second Industrial Revolution.

The beauty of the IC has not escaped the contemporary graphic designer. This intricate circuitry with its intriguing pattern of lines, often tapering from thick to thin, is the physical embodiment of unseen power. It is appropriate then that these linear compositions, seemingly random yet actually highly structured, have become the visual metaphor for an entire industry. The function of the IC is indeed the basis of all high technology: the input, assimilation, and output of data. Since many high tech companies utilize this process in their products or services, it is not uncommon for their trademarks to incorporate multiple lines mimicking the ICs upon which they rely.

Technology companies of the 1980s embrace the multiple line much the same way as industrial companies of the 1930s utilized the lightning bolt. The zig-zag image of lightning streaking across a corporate mark was the designer's metaphor for electricity, the unseen yet driving force of industry after

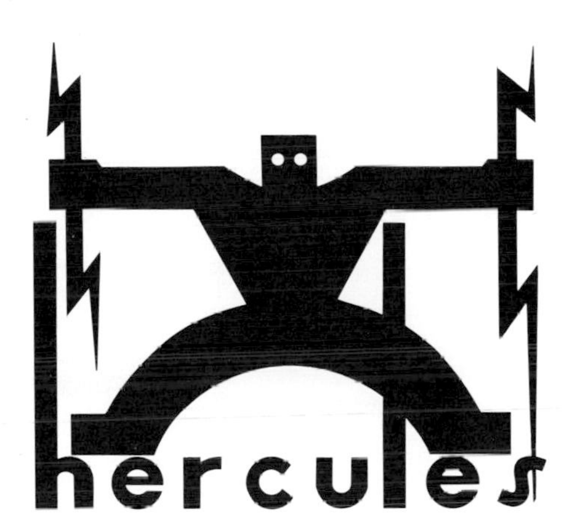

The lightning bolt was a popular icon for trademark designers in the 1930s who were seeking a visual metaphor for the power of American industry after the Great Depression. Today, the multiple line is commonly used as a symbol for the dynamicism of our high technological era.

the Great Depression. A half century later the line, slim or tapering, has become the symbol for our electronic age of information processing. The high speed, highly miniaturized and highly powerful transfer of information through the lines of electronic circuits is truly the "high" of high technology.

Other imagery derived from technology is assimilated into high tech identity. The I and O binary code of computers, the step form rastering from video displays, the oscilloscope wavelength, and the CRT screen and floppy disk are some of the icons employed.

Intriguingly, a number of companies utilize uncommon and unexpected objects to represent themselves, Apple Computer being the best known example. Software publishers in particular often prefer images from nature – the antithesis of high tech – perhaps to mollify the anxious consumer. By projecting a bit of familiarity onto an unfamiliar (and likewise intimidating) product, these corporations wrap the high tech experience in an aura of down-home friendliness. Inserting a floppy disk into a personal computer is as simple as

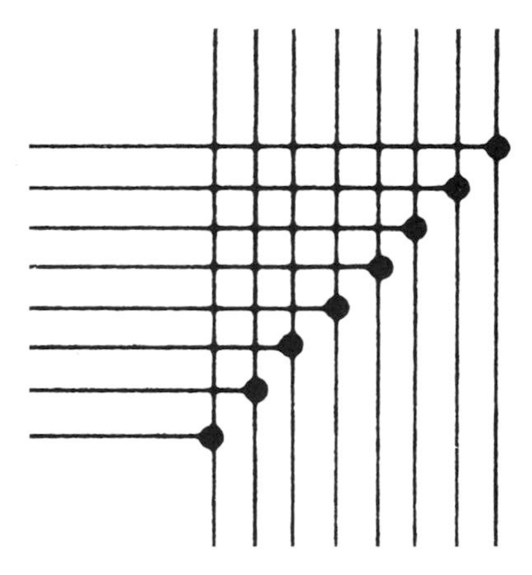

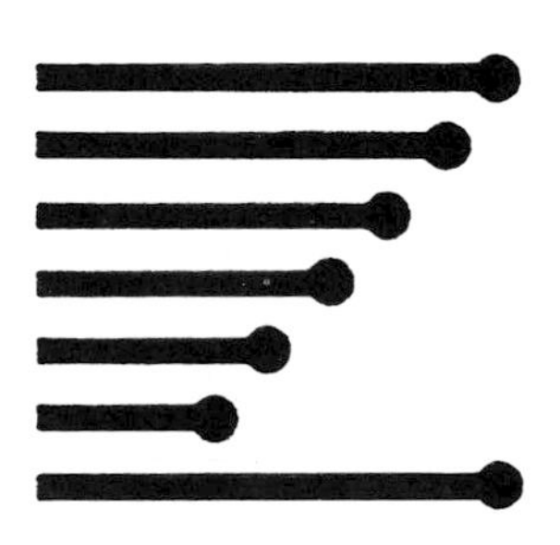

Detail of an IC design; on the right is the trademark for Zyvex Corporation

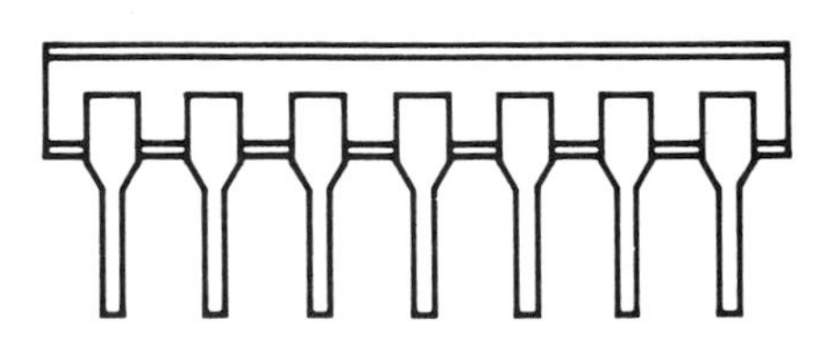

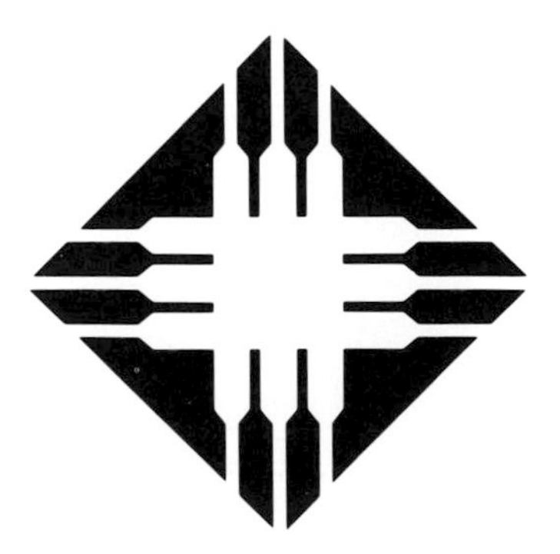

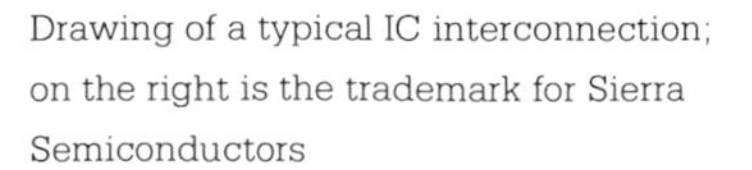

Drawing of a typical IC interconnection; on the right is the trademark for Sierra Semiconductors

plucking a peach from a tree in the backyard, and software publishers certainly have little to lose by suggesting such an analogy. Although it may be risky to predict future marketing strategies, it is probable that companies targeting the home market will increasingly embrace a more casual identity than those selling primarily to business clientele.

The evolution of the AT&T logo illustrates the shift in graphic design towards more abstract (technological) forms and away from object-oriented (industrial) symbolism. The folksy bell is eventually reduced to a simplified, bell-like image which borders on the non-objective. No longer solely a telephone industry, the current AT&T mark emphasizes the corporation's involvement in telecommunications and computers. Although seemingly computer-generated, the logo was actually developed through a series of over 2000 hand rendered roughs. It is perhaps ironic that one of the most memorable symbols of modern technology was conceived with a pencil!

Typographic designs are frequently utilized by high tech corporations. An

The first Apple Computer trademark designed by Ronald G. Wayne imparted a distinct medieval character.
Rob Janoff's design for the current logo mixes metaphors: the apple is a symbol for intellect while the bite removed suggests the eating of the forbidden fruit

early example is the IBM mark of 1945 which featured the words "International Business Machines" forming a globe, emphasizing the international aspect of the company; in 1955 this unassuming design gave way to Paul Rand's bold, square serif logotype. Running the gamut from digital letterforms to handwritten script, logotypes stress name recognition over visual metaphor. The designer's challenge is to create interest with given, and often unwieldy typographic combinations. Not often easy to do, these designs underscore the difficulty of manipulating type without succumbing to cliché.

Indeed, the creation of corporate identity which symbolizes modern technology yet appears fresh and memorable in the eyes of the consumer will become an increasingly formidable task. Logos which go beyond the commonplace will require a level of research and experimentation which is currently only moderately apparent. The dangers of redundancy in such a highly competitive industry cannot be ignored. Perhaps the solution to this problem will come from the industry itself: like it or not the computer will ultimately play a much greater role in determining the look of the next generation of high technology trademarks. And more people will be cooking with microwaves.

1889

1900

1921

1939

1964

1969

As the Bell Telephone logo evolved, it became simpler and less objective. The current AT&T mark is totally abstract, suggesting more the company's involvement with global telecommunications than its origins in the telephone industry.

IBM

The original IBM logotype (circa 1945) and the 1955 redesign by Paul Rand

1 **Nippon Kogaku, K.K.**

Tokyo, Japan
Nikon cameras
Design: Yusaku Kamekura

1

2	**MultiTech Systems**	New Brighton, Minnesota *Data communications equipment* Design: P. Scott Makela
3	**Advanced Micro Devices**	Sunnyvale, California *Integrated circuits* Design: Lawrence Bender & Associates
4	**Madic Corporation**	Santa Clara, California *Corporate information system software* Design: Sharp Communications
5	**Binary Technology**	Meriden, New Hampshire *Single board computers* Design: Rose Russo Crooker
6	**Analog Devices**	Norwood, Massachusetts *Electronic components*
7	**Precision Echo**	Santa Clara, California *Instrumentation tape recording devices* Design: Carolee Robbins-Gonzalez

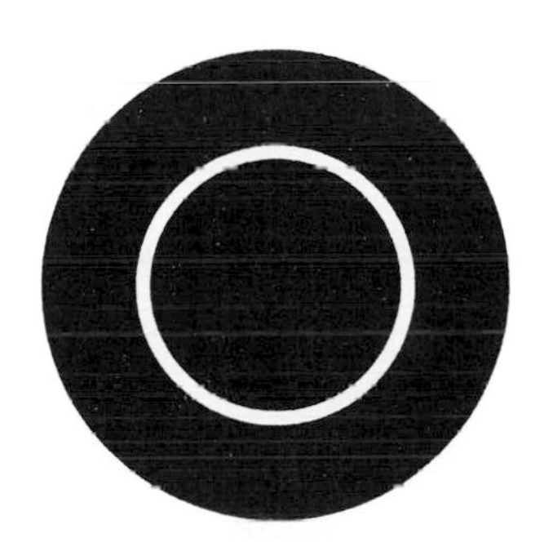

2

3

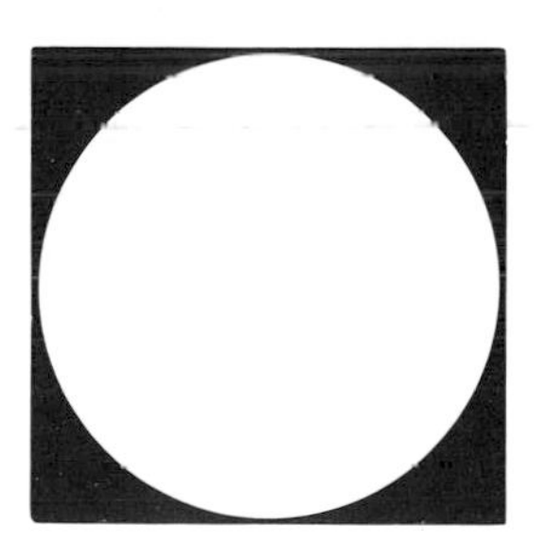

4

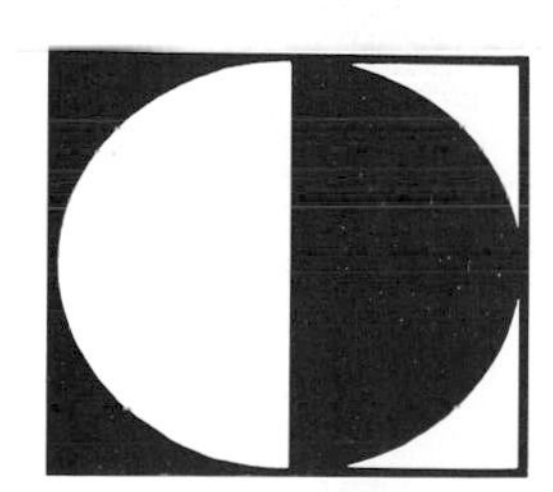

5

6

7

8	**Polytech GmbH**	Karlsruhe, West Germany *Lasers*
9	**Hughes Helicopters**	Culver City, California *Helicopters*
10	**Siltec**	Menlo Park, California *Silicon wafer manufacturing equipment* Design: Frank A. Rodriquez
11	**Creative Output**	Milford, Connecticut *Management systems software & consulting* Design: Dan Reisinger
12	**Sonar Corporation**	Cherry Hill, New Jersey *Electronic equipment*
13	**Keicho Ltd.**	Tokyo, Japan *Computer software* Design: Kazumasa Nagai

8

0

10

11

12

13

14 **Communications Industry Association of Japan**

Tokyo, Japan
Telecom-technology exhibitions
Design: Kazmuasa Nagai

15	**Hitachi Seiko Ltd.**	Kanagawa-Ken, Japan *Semiconductors, computers & peripherals* Design: Namihei Odaira
16	**Integra Microwave**	Santa Clara, California *Microwave test instruments* Design: Werner Schuerch
17	**Technical Optics Ltd.**	Isle of Man, British Isles *Laser optics & interferometers*
18	**Compumotor Corporation**	Petaluma, California *Motion control components*
19	**Century Software**	Sandy, Utah *Communications software*
20	**Communication Transistor Corporation**	San Carlos, California *Microwave transistors* Design: Lawrence Bender & Associates

15

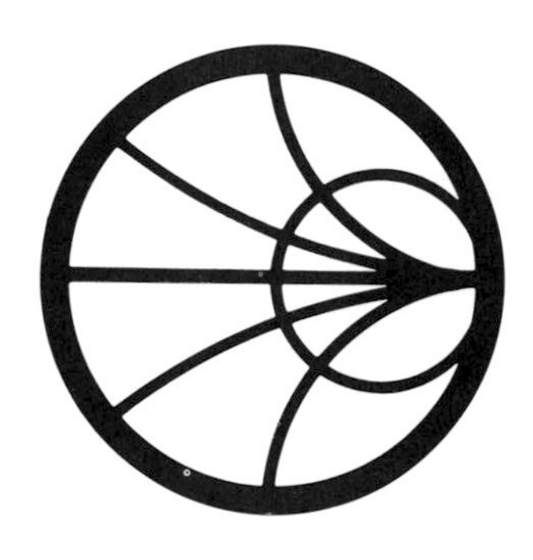

16

17

18

19

20

No.	Company	Location / Business / Design
21	**Sanyo Corporation**	Moonachie, New Jersey *Electronic products*
22	**Royal Micrographics**	San Francisco, California *Storage & retrieval systems* Design: Primo Angeli Graphics
23	**Computer Cognition**	National City, California *Accounting software for microcomputers* Design: Richard McWhorter
24	**Applied Solar Energy Corporation**	City of Industry, California *Solar cells & cell assemblies* Design: Robert, Miles, Runyan & Associates
25	**Inmos, Ltd.**	Bristol, England *Integrated circuits & computer software* Design: Don Burston & Associates
26	**Aeon Systems**	Albuquerque, New Mexico *Electronic data acquisition equipment* Design: David Kunzelman

21

22

23

24

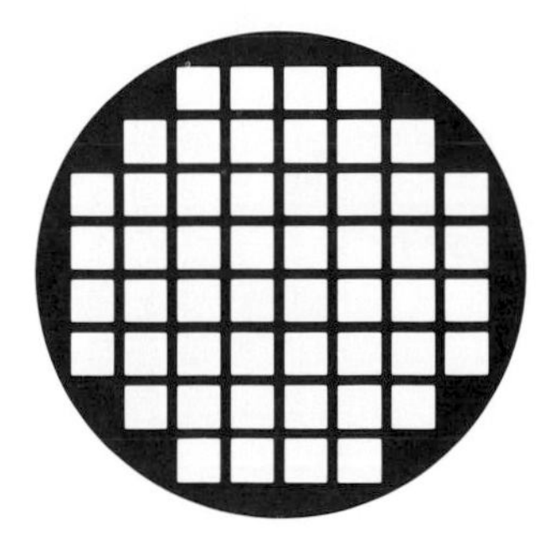

25

26

27 **Universal Semiconductor**

San Jose, California
Semiconductors
Design: George Opperman

28	**Computer Business Solutions**	Marietta, Georgia *Computer Peripherals* Design: David Associates
29	**Huntley Knight**	Los Gatos, California *Electronic/optical measuring equipment* Design: Gruye Associates
30	**Fiat Termomeccania S.p.A.**	Torino, Italy *Gas turbines & cogeneration power stations*
31	**Minami Measuring Apparatus Ltd.**	Tokyo, Japan *Measuring instruments* Design: Yusaku Kamekura
32	**Microscience International**	Mountain View, California *Winchester disk drives*
33	**Thermo-Cell Southeast**	Atlanta, Georgia *Hybrid polyisocyanurate foam (product logo)* Design: David Laufer Associates

28

29

30

31

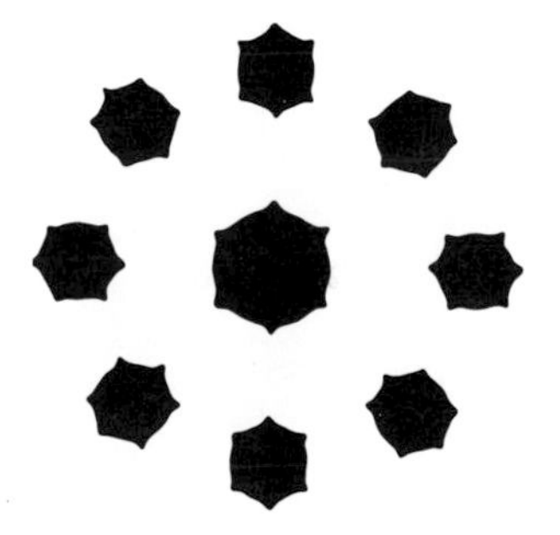

32

33

34	**Paragon Software**	Boca Raton, Florida *Software development and marketing* Design: The Phoenix Group
35	**BASF Systems Corporation**	Bedford, Massachusetts *Magnetic media* Design: In-house
36	**Gates Learjet Corporation**	Tucson, Arizona *Business jet aircraft* Design: In-house
37	**Farnell Instruments Ltd.**	West Yorkshire, England *Stabilized power equipment* Design: Wilfred Hall Associates
38	**Winegard Company**	Burlington, Iowa *Satellite TV equipment*
39	**Antenna Technology Corporation**	Scottsdale, Arizona *Microwave components*

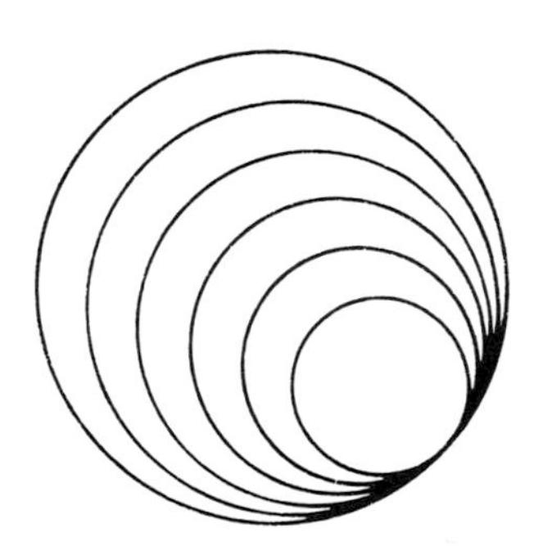

34

35

36

37

38

39

40 **Met One**

San Mateo, California
Windspeed indication equipment
Design: Michael Vanderbyl

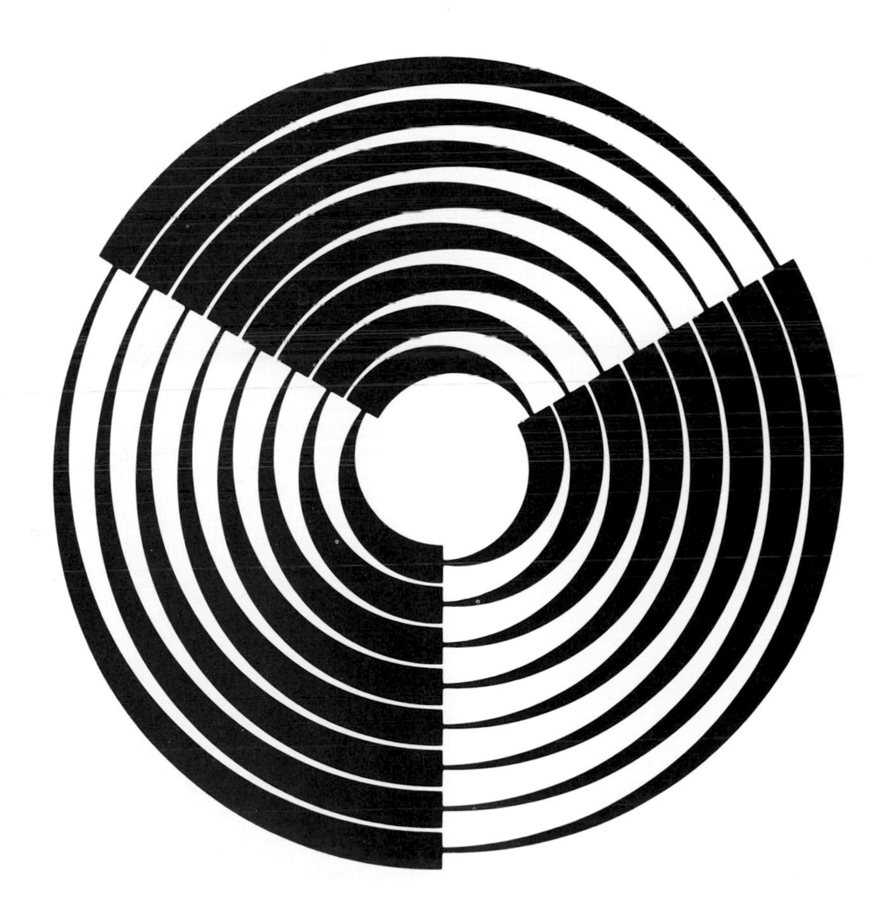

41	**INTERMEC**	Lynwood, Washington *Industrial bar code scanners & readers*
42	**Digital Research**	Pacific Grove, California *Computer software*
43	**Bowmar Instrument Corporation**	Clearwater, Florida *Computer keyboards & displays* Design: Stu Maws
44	**Beseler**	Florham Park, New Jersey *Camera system for computer graphics displays* Design: Kalmar Ad Marketing
45	**Eastern Software Products**	Alexandria, Virginia *Software systems* Design: Dorothea C. McGay
46	**Rohr Industries**	Chula Vista, California *Commercial & military aircraft components*

41

42

43

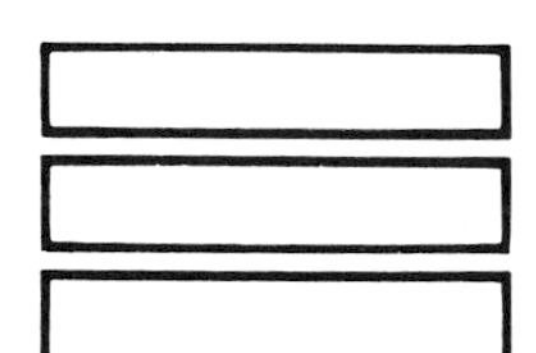

44

45

46

47	**Advanced Robotics Corporation**	Columbus, Ohio *Computer controlled robotic work cells* Design: Lord, Sullivan & Yoder
48	**Romcor Design**	Cupertino, California *Computer software/hardware* Design: David Kelley Design
49	**Divelbiss Corporation**	Fredricktown, Ohio *Electronic products engineering*
50	**Apstek Incorporated**	Dallas, Texas *Computer memory expansion boards*
51	**Data 3 Systems**	Santa Rosa, California *Manufacturing software for the IBM PC* Design: In-house
52	**PCT Technologies**	Ann Arbor, Michigan *Multi-terminal processing system for IBM PCs* Design: In-house

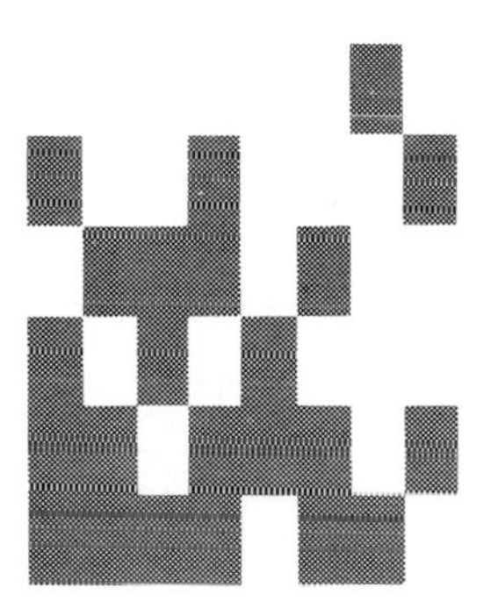

47

48

49

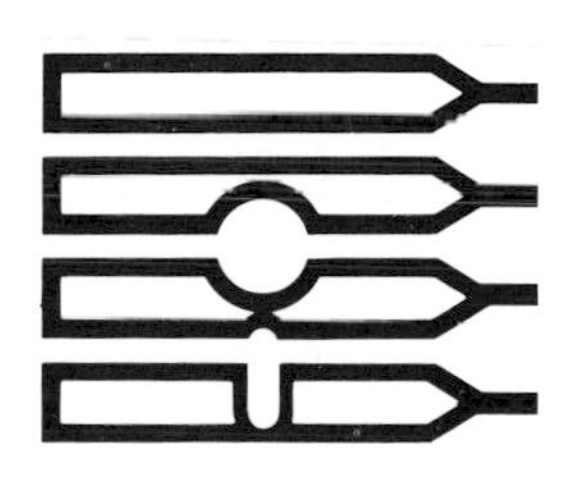

50

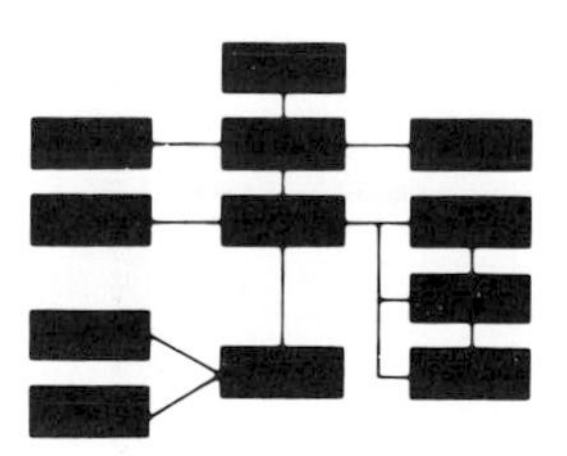

51

52

53 **Onset**

Palo Alto, California
Seed capital
Design: David Kelley Design

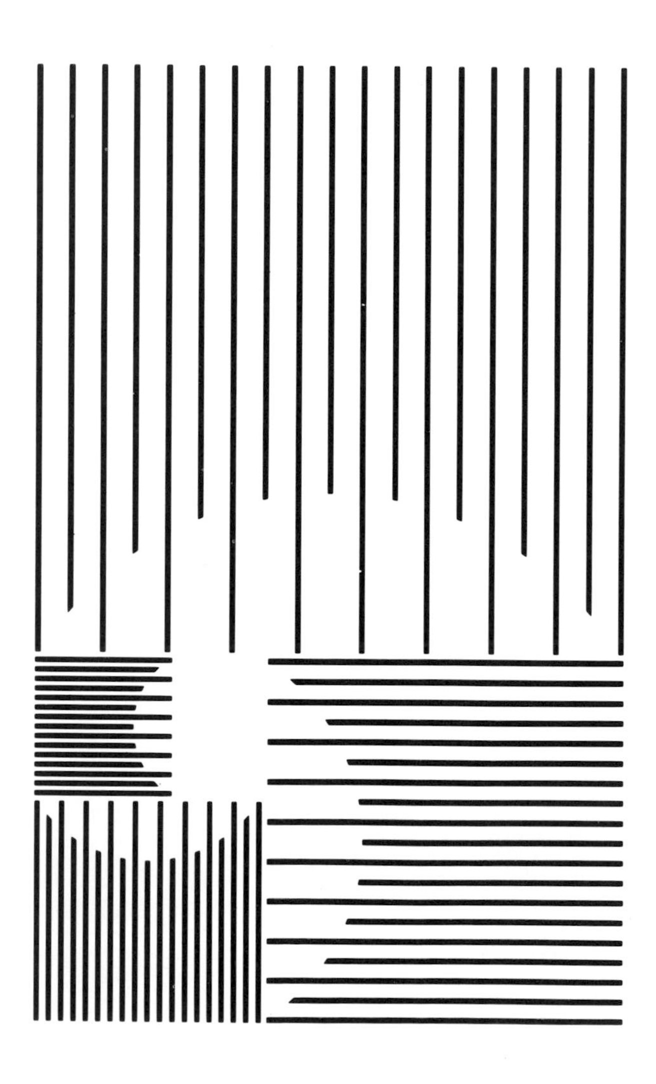

54	**SYSTAT Incorporated**	Evanston, Illinois *Statistics Software* Design: Howard Blake & Company
55	**Presentation Technologies**	Palo Alto, California *PC-based slide generating equipment* Design: David Kelley Design
56	**CertiFLEX Systems**	Dallas, Texas *Accounting software* Design: Dennis Crudden
57	**National Product Marketing**	Atlanta, Georgia *Computer interfaces* Design: Weir and Pfeiffer
58	**Slicer Computers**	Minneapolis, Minnesota *Computer systems* Design: Rich Harms
59	**Sorcim/IUS**	San Jose, California *Management software*

54

55

56

57

58

59

60	**Star Technologies**	Portland, Oregon *High-speed array processor computers* Design: Rudi Milpacher
61	**Digi-Data Corporation**	Jessup, Maryland *Computer peripherals* Design: Tony DiPinto
62	**Micro Business Applications**	Burnsville, Minnesota *Computer software* Design: Woody Pirtle
63	**Terminal Data Corporation**	Woodland Hills, California *Computer controlled micrographics systems* Design: Tom Rigsby
64	**Storm Products Company**	Hinsdale, Illinois *Microwave & fiber optic cables*
65	**On-Line Software International**	Fort Lee, New Jersey *Computer software* Design: Joanna Sluja

60

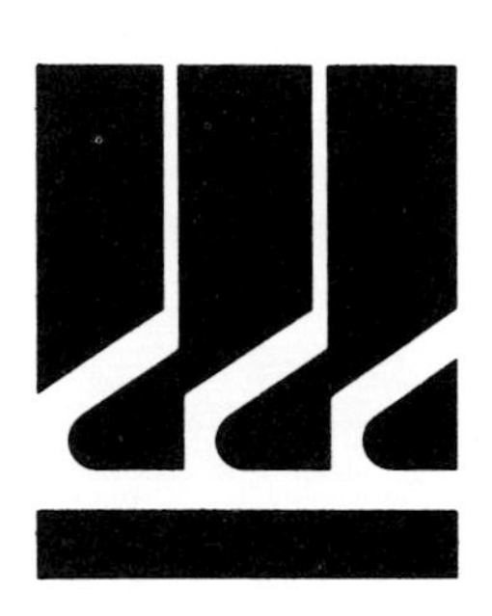

61

62

63

64

65

66 **Ultratech Stepper**

Santa Clara, California
Photolithography equipment
Design: Mark Anderson Design

67	**Polygon Associates**	St. Louis, Missouri *Communication software* Design: In-house
68	**Primavera Systems**	Bala Cynwyd, Pennsylvania *Software for project management & control* Design: In-house
69	**J.C. Schumacher Company**	Oceanside, California *Semiconductor materials research & development*
70	**Thermo-Cell Southeast**	Atlanta, Georgia *Hybrid polyisocynurate foam (product logo)* Design: David Laufer Associates
71	**Checkpoint Data**	Santa Ana, California *Computer consulting* Design: Hedi Yamada
72	**Labsystems**	Chicago, Illinois *Laboratory equipment* Design: Anderson & Lembke

67

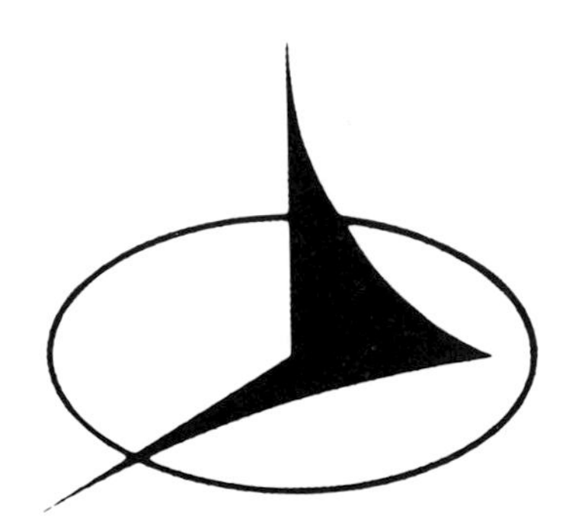

68

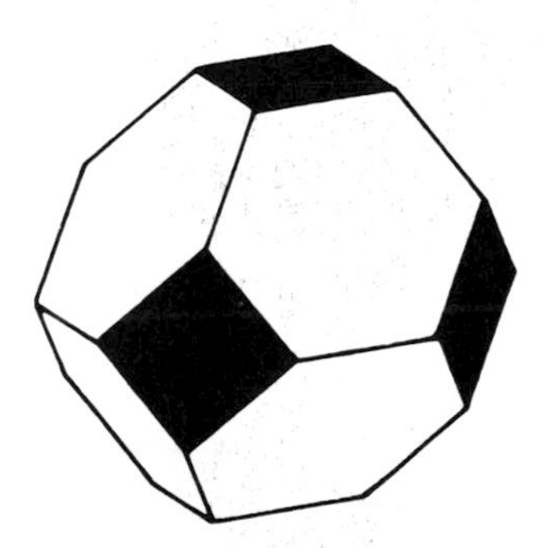

69

70

71

72

73	**Matrix Instruments**	Orangeburg, New York *Color recorders for computer-generated graphics* Design: In-house
74	**Antech Incorporated**	Roswell, Georgia *Architectural & engineering software* Design: Rod Hoffman
75	**Comchek International**	Scottsdale, Arizona *Computer network testing equipment* Design: Redding & Associates
76	**Bond-Tech**	Englewood, Ohio *Business software*
77	**Pac Tec Corporation**	Philadelphia, Pennsylvania *Electronic enclosures* Design: Peter A. Peroni
78	**Software Building Blocks**	Ithaca, New York *Language software & development tools* Design: Laurie H. Moskow

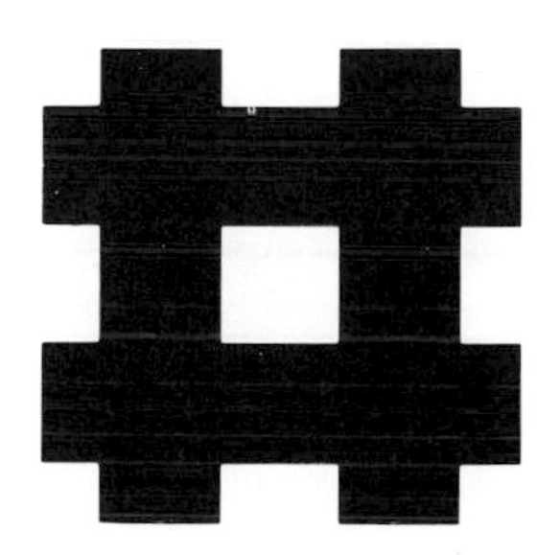

73

74

75

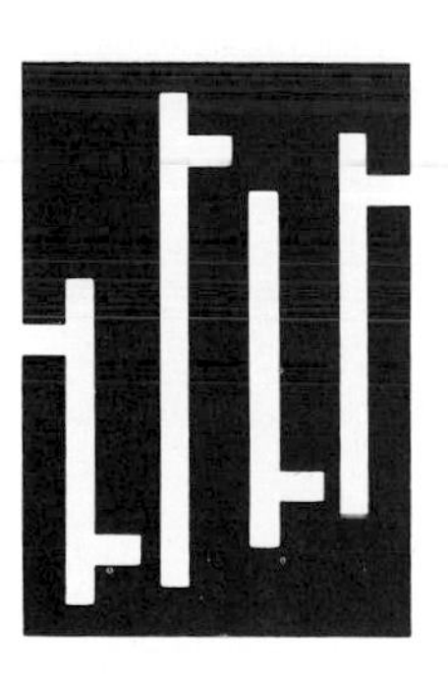

76

77

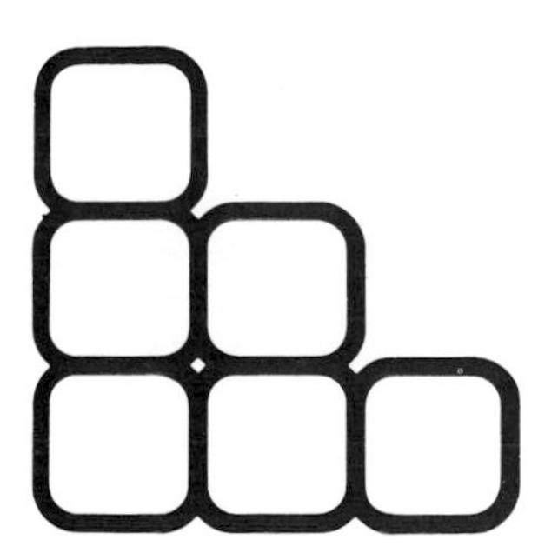

78

 Quadram Corporation

Norcross, Georgia
Computer peripherals
Design: David Giles

No.	Company	Location / Business / Design
80	**Blaise Computing**	Berkeley, California *Computer peripherals* Design: Susan Thomson
81	**Sperry Corporation**	Blue Bell, Pennsylvania *Electronics-based systems* Design: Robert Pucci
82	**Precision Visuals**	Boulder, Colorado *Interactive computer graphics systems* Design: Bryan Gough
83	**Systems Strategies**	New York, New York *Professional software consulting*
84	**The SIMCO Company**	Hartfield, Pennsylvania *Static electricity control products* Design: Tech-Ad
85	**GIXI Incorporated**	Edina, Minnesota *Computer graphics displays*

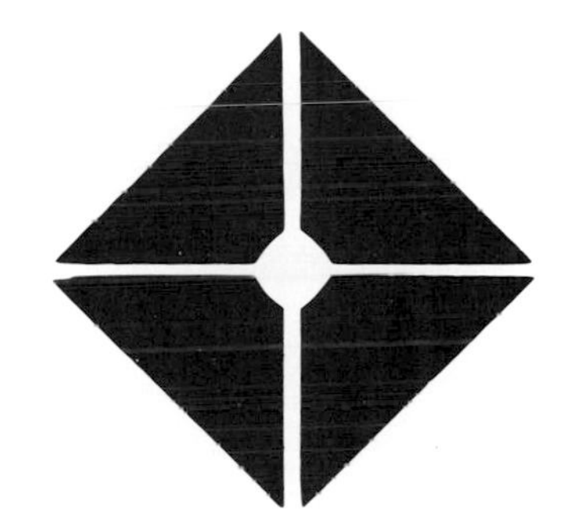

80

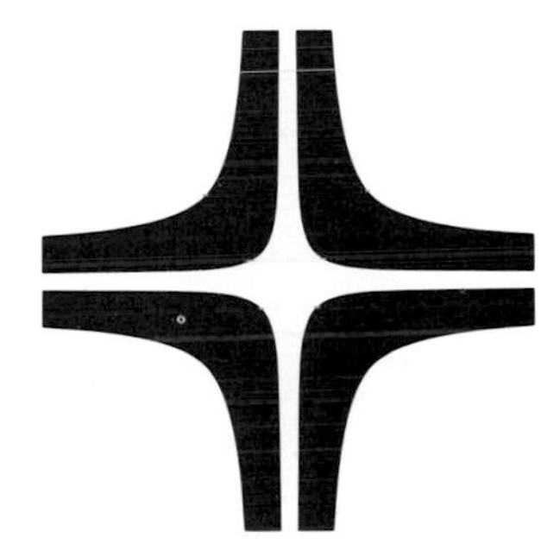

81

82

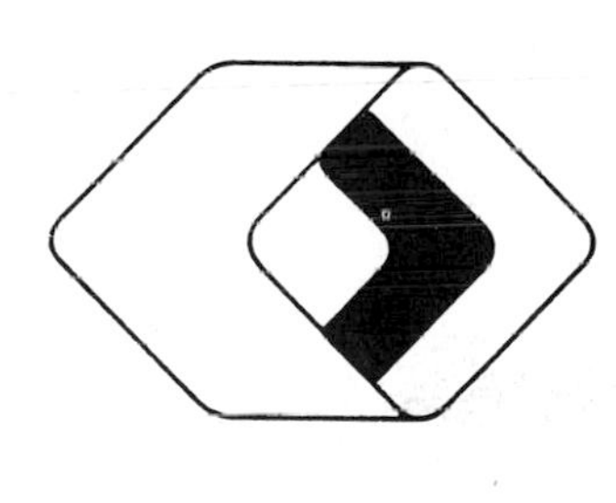

83

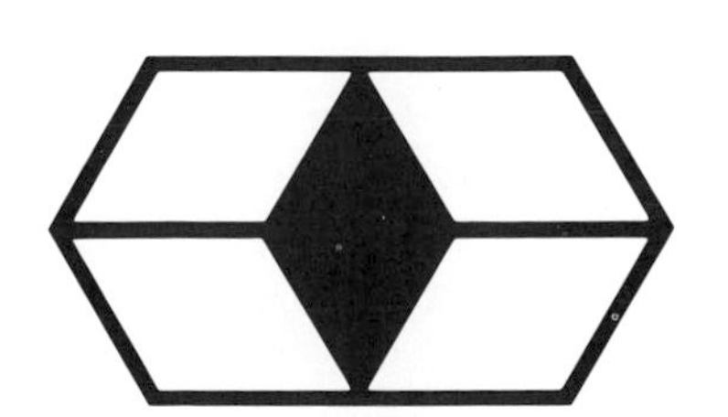

84

85

86	**Jodon Laser**	Ann Arbor, Michigan *Helium neon lasers & holographic equipment* Design: In-house
87	**Borg Warner Corporation**	Chicago, Illinois *Environmental control devices* Design: Anspach Grossman Portugal
88	**Uchihashi Metal Industrial Company**	Osaka, Japan *Thermal protectors for electronic circuits*
89	**E+E DataComm**	San Jose, California *Data communications* Design: Steinhilber, Deutsch & Gard
90	**Intelligent Technologies**	Palo Alto, California *Integrated hardware/software packages*
91	**Interactive Software**	Mountain View, California *Productivity software for microcomputers* Design: Primo Angeli Graphics

86

87

88

89

90

91

92 **SensorMedics Corporation**

Anaheim, California
Electronic medical instrumentation
Design: Cochrane Chase Livingston & Co.

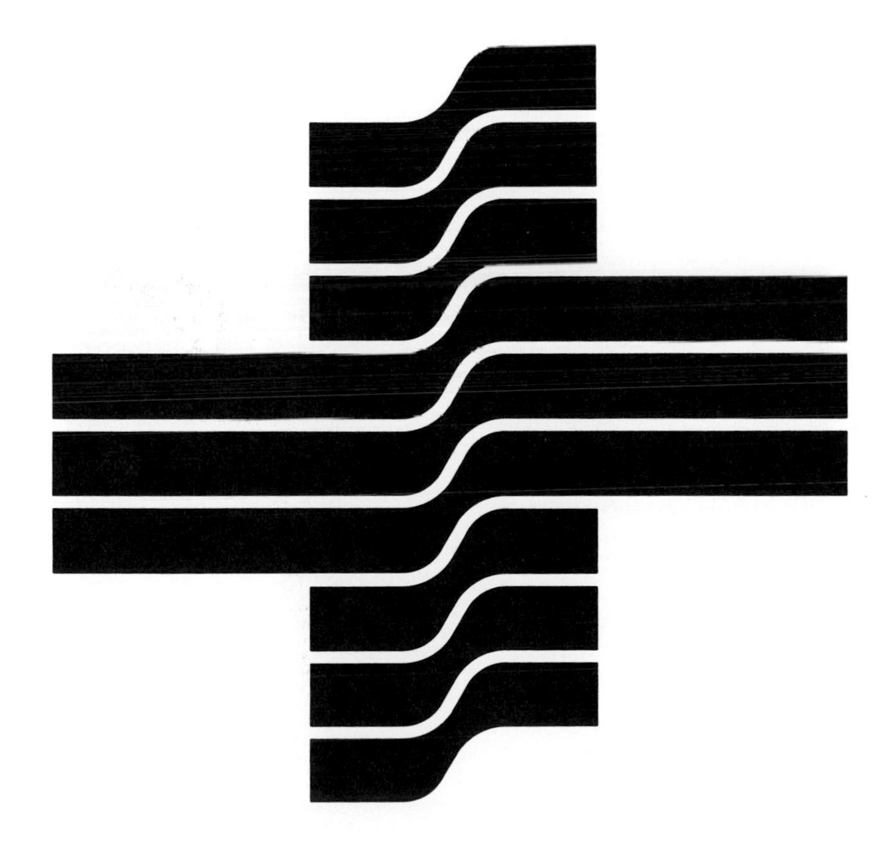

93	**AT&T**	New York, New York *Telecommunications* Design: Saul Bass/Herb Yager Associates
94	**Assimilation Incorporated**	Los Gatos, California *Software for the Apple Macintosh* Design: Ellen Gartland
95	**Ensys**	Houston, Texas *Computer analysis for oil companies* Design: Chris Hill
96	**Phact Associates**	New York, New York *Database management software* Design: Margarita Carreras
97	**Yamagiwa Corporation**	Tokyo, Japan *Lighting devices* Design: Yusaku Kamekura
98	**Yamagiwa Kosakusho**	Tokyo, Japan *Materials for lighting manufacture* Design: Yusaku Kamekura

93

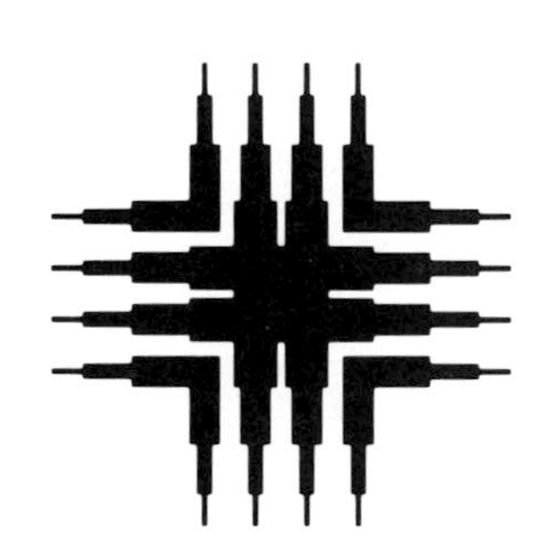

94

95

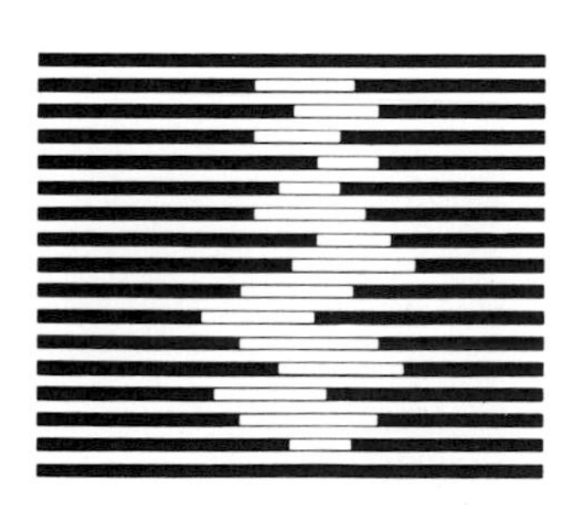

96

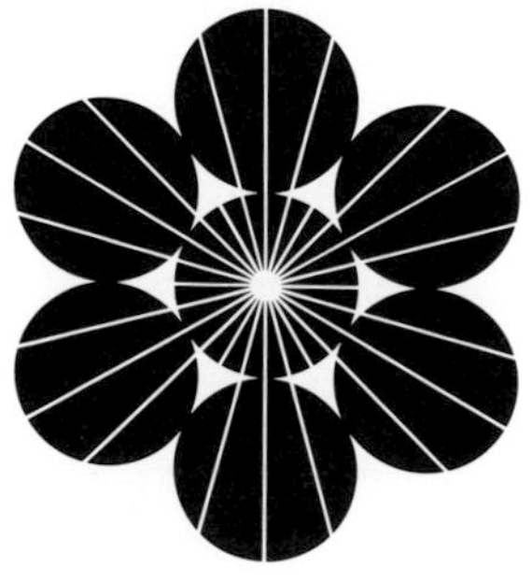

97

98

No.	Company	Location / Business / Design
99	**Electronic Arts**	San Mateo, California *Computer software* Design: Steinhilber, Deutsch & Gard
100	**Sierra Semiconductors**	Sunnyvale, California *Semiconductors* Design: Suzanne Redding
101	**Delta Technology**	Eau Claire, Wisconsin *Software for the IBM PC* Design: Craig Smith
102	**Datacopy Corporation**	Mountain View, California *Electronic digitizing cameras* Design: Hovey-Kelley Design
103	**Nantucket Corporation**	Malibu, California *Computer hardware/software* Design: Guzman/Gerrie Advertising
104	**Adax Incorporated**	Atlanta, Georgia *Telephone/computer interface systems* Design: Beth Tippett

99

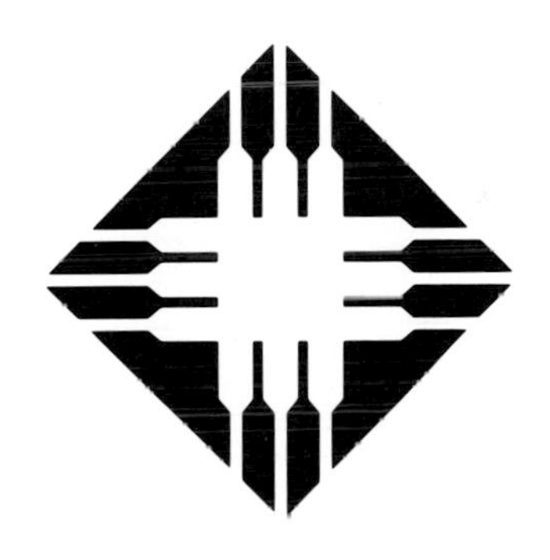

100

101

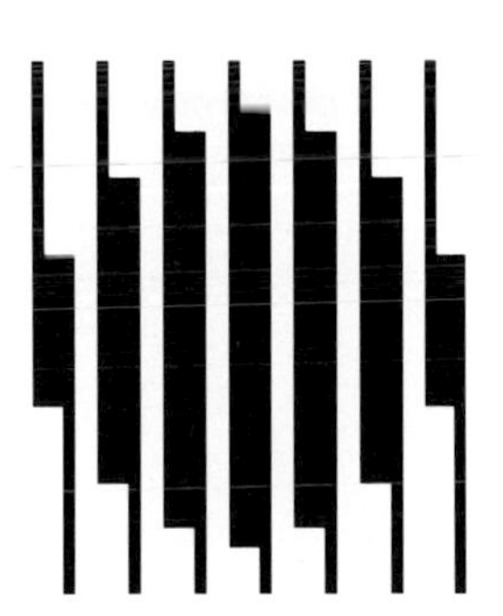

102

103

104

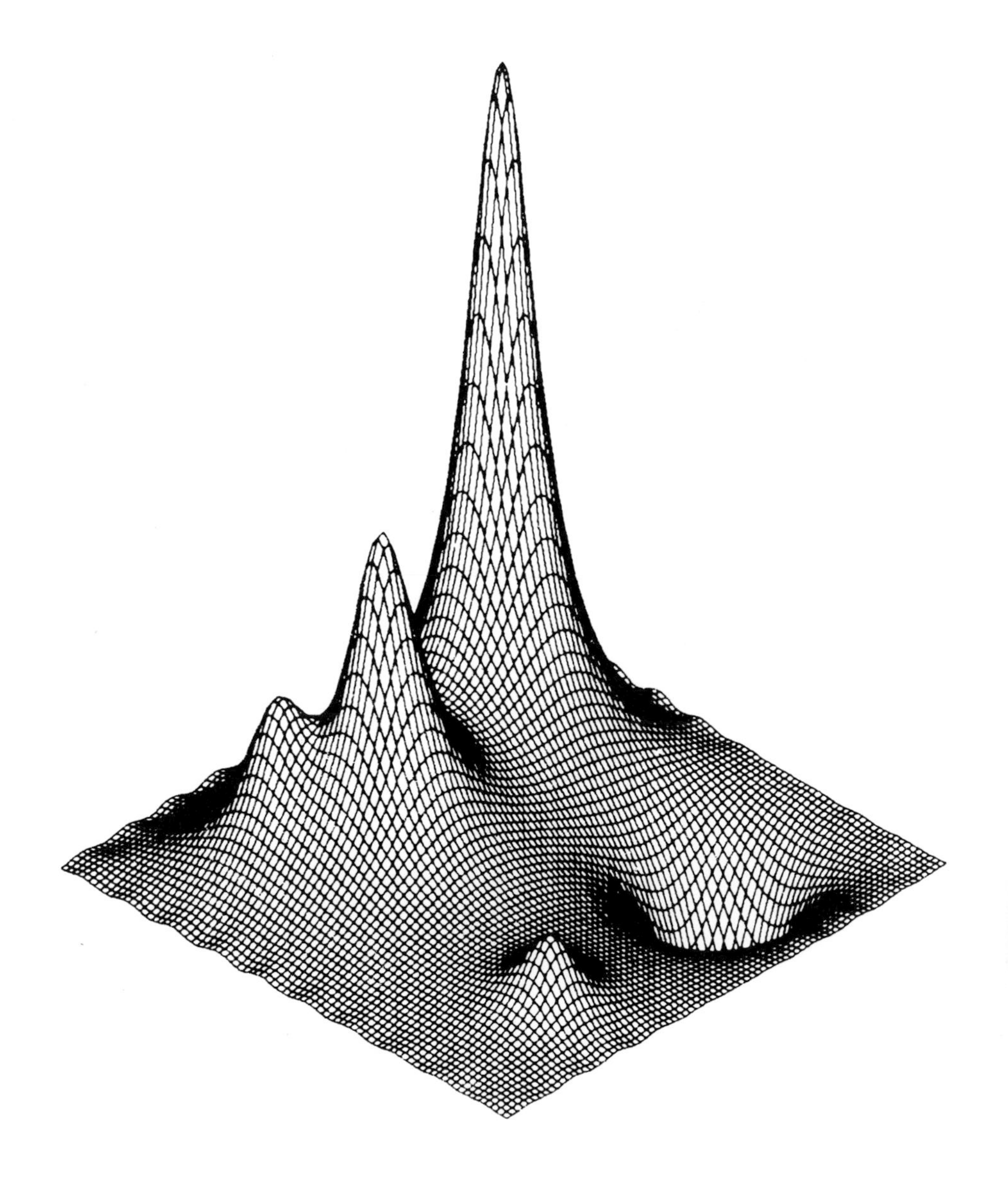

106	**James B. Lansing**	Los Angeles, California *Audio equipment* Design: Jerome Gould
107	**Southern California Gas Company**	Los Angeles, California *Electronic symbol for gas as energy source* Design: Jerome Gould
108	**Glassman High Voltage**	Whitehouse Station, New Jersey *Electronic equipment*
109	**Partlow Communications**	New Hartford, New York *Electronic controls* Design: Doug Cleminshaw
110	**Electromagnetic Sciences**	Norcross, Georgia *Electromagnetic equipment*
111	**Raytel Medical Imaging**	Campbell, California *Picture archives and communication systems* Design: Richard Bader

106

107

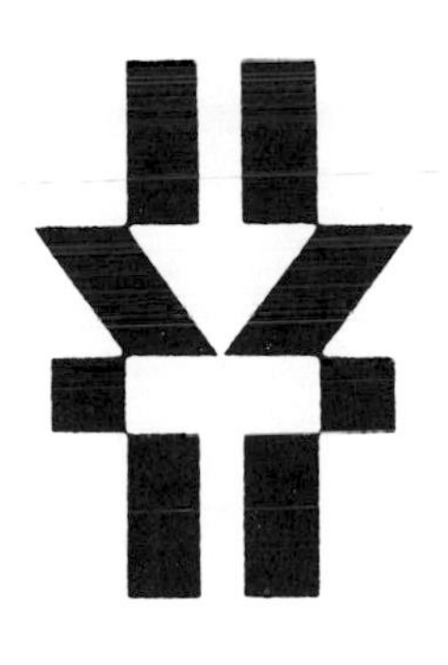

108

109

110

111

112	**Cadnetix Corporation**	Boulder, Colorado *Turnkey workstations for electronic design* Design: Hanlon Design Group, Ltd.
113	**Canstar Communications**	Scarborough, Ontario, Canada *Fiber optic cables & couplers* Design: the Mobius strip
114	**The Software Link**	Atlanta, Georgia *Software for the IBM PC* Design: Rod Roark
115	**Concord Data Systems**	Waltham, Massachusetts *Data communications equipment* Design: Impact Marketing and Communications
116	**BYTCOM**	San Rafael, California *Modems* Design: Owens-Lutter
117	**Ungermann-Bass**	Santa Clara, California *Computer network systems* Design: Lawrence Bender & Associates

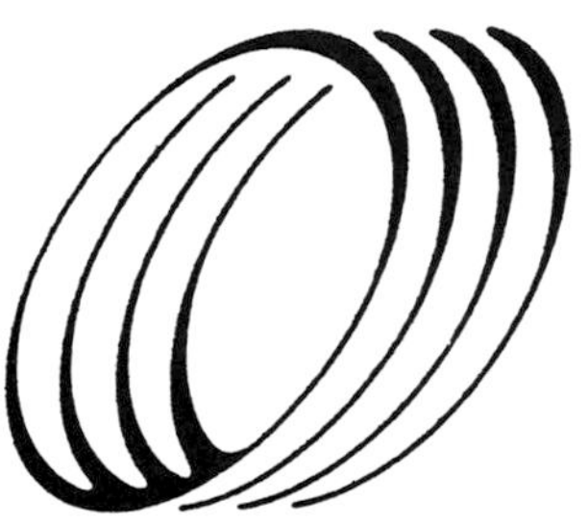

112

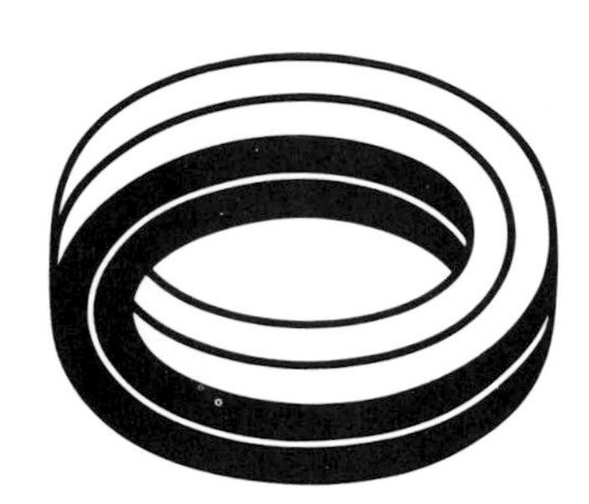

113

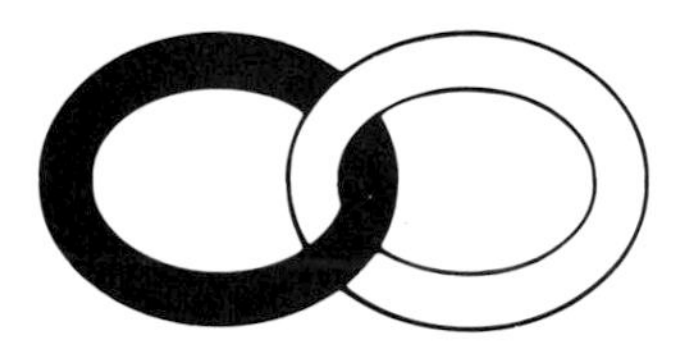

114

115

116

117

Berkeley, California
Engineering analysis software

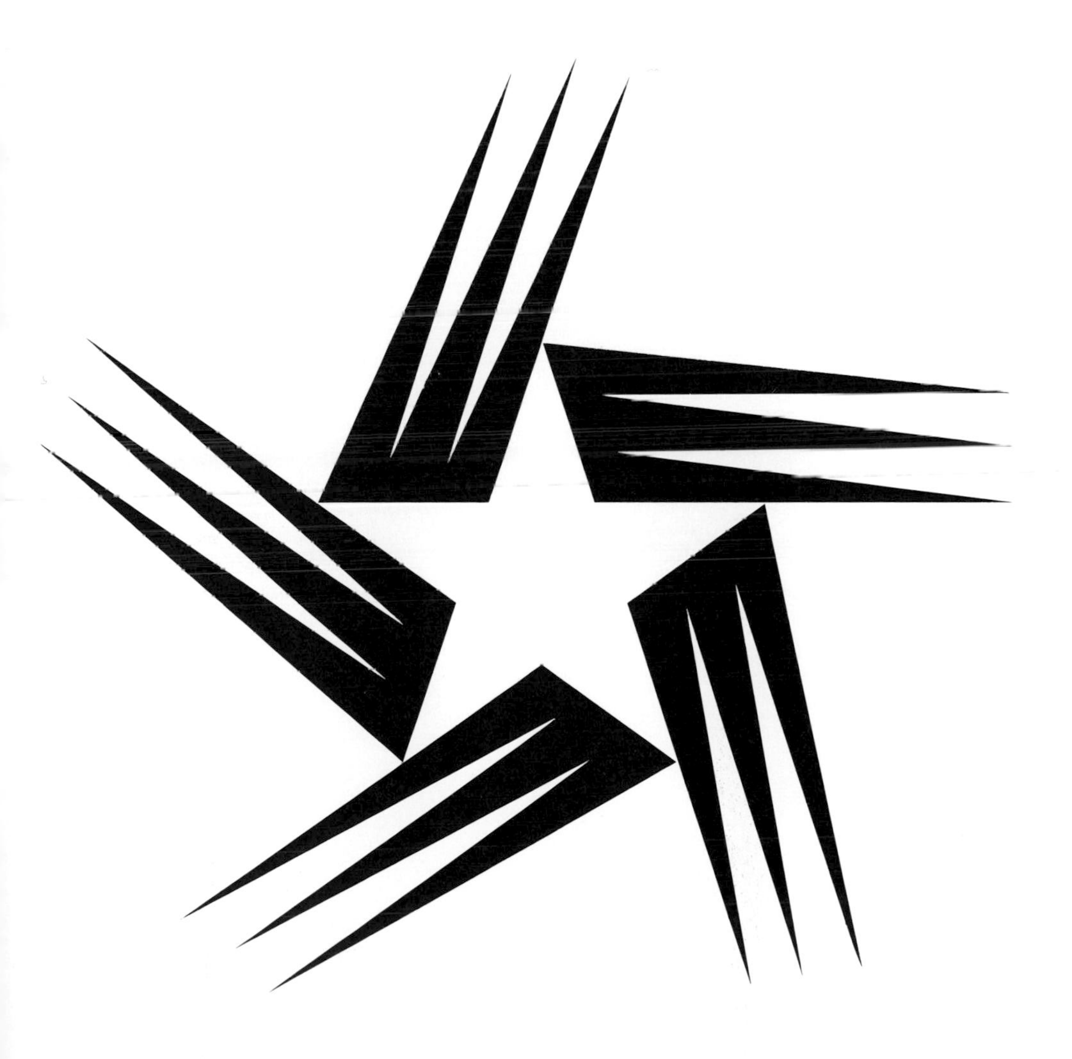

119	**Diversified Computer Corporation**	Seattle, Washington *Minicomputer turnkey systems* Design: Lawrence Bender & Associates
120	**General Cable Company**	Greenwich, Connecticut *Wire for electronic equipment* Design: Tom Burns Associates
121	**Yamaha International**	Buena Park, California *Electronic music equipment* Design: Nippon Gakki Company, Ltd.
122	**EG&G Incorporated**	Wellesley, Massachusetts *Scientific instruments* Design: Hammond Design Associates
123	**Allied Corporation**	Morristown, New Jersey *Aerospace & chemical products* Design: Lippincott & Margulies
124	**ACCUCOM Data Network**	Tigard, Oregon *Typesetting & laser production software* Design: Berni Moor

119

120

121

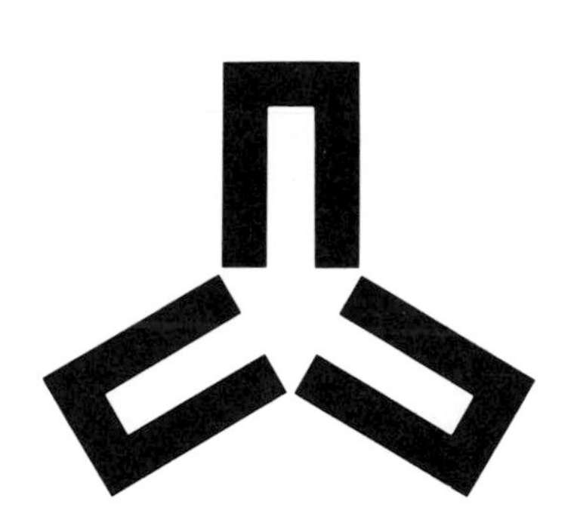

122

123

124

125	**Silicon Graphics**	Mountain View, California *Computer graphics hardware/software* Design: Scott Kim
126	**Applied Software Technology**	Los Gatos, California *Database management business software*
127	**TDK Corporation**	Tokyo, Japan *Audio components* Design: Yusaku Kamekura
128	**Advanced Crystal Sciences**	San Jose, California *Semiconductor fabrication equipment* Design: In-house
129	**CYBOTECH Corporation**	Indianapolis, Indiana *Industrial robot systems*
130	**Harvard Software**	Littletown, Massachusetts *Management productivity software* Design: James Rue Design

125

126

127

128

129

130

131 **CERICOR**

Salt Lake City, Utah
Electronic design automation systems
Design: Lucas R. Vissar

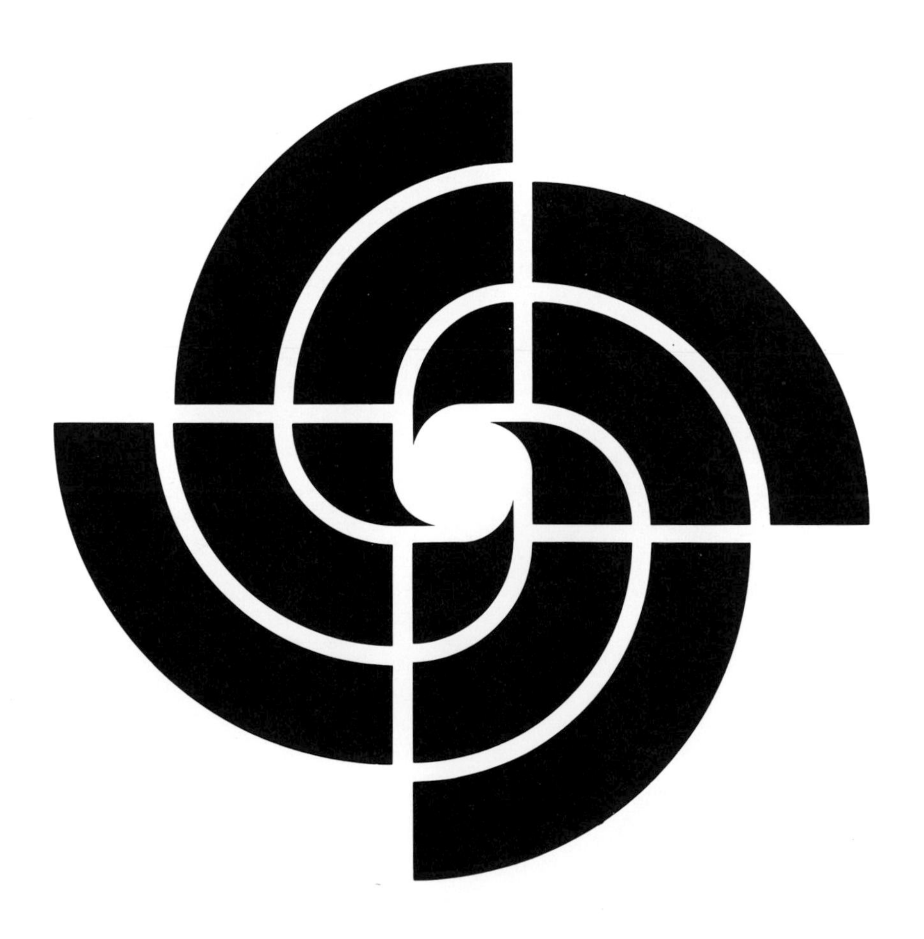

132	**Consulting Services Associates**	Los Gatos, California *Software consulting* Design: Patrick Mountain Design
133	**Menlo Corporation**	Santa Clara, California *Database access software* Design: Stephen Jacobs Fulton & Green
134	**Ven-Tel**	Santa Clara, California *Modems & line drivers*
135	**Taiyo Machine Industry Company Ltd.**	Tokyo, Japan *Electronic equipment* Design: Yusaku Kamekura
136	**Softstyle Incorporated**	Honolulu, Hawaii *Computer software development* Design: Turner and deVries
137	**Silicon Compilers**	Pasadena, California *Software for integrated circuit design* Design: Lawrence Bender & Associates

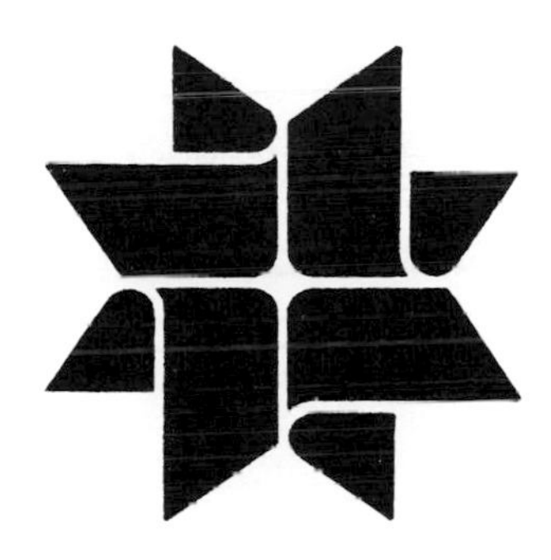

132

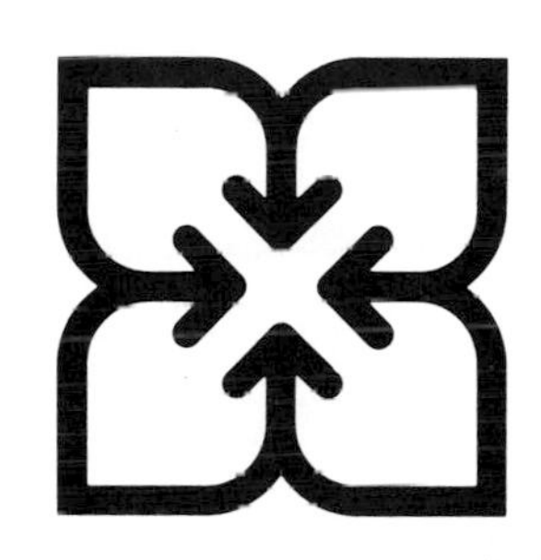

133

134

135

136

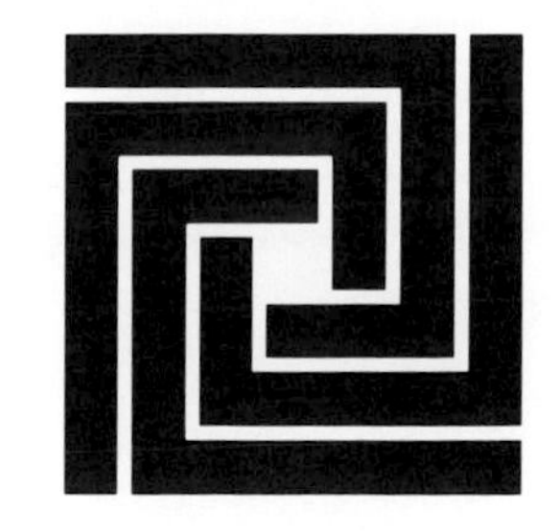

137

138	**Comnet**	Washington, D.C. *Computer timesharing services*
139	**Systems Plus**	Palo Alto, California *Business software*
140	**Teleautograph**	Los Angeles, California *Two way communication systems* Design: Jerome Gould
141	**Digital Systems Corporation**	Walkersville, Maryland *Minicomputers & disk drive controllers*
142	**Fusion Systems Corporation**	Rockville, Maryland *Semiconductors*
143	**WESPERCORP**	Tustin, California *Mass memory and peripheral control devices* Design: Dick Sylvester

138

139

140

141

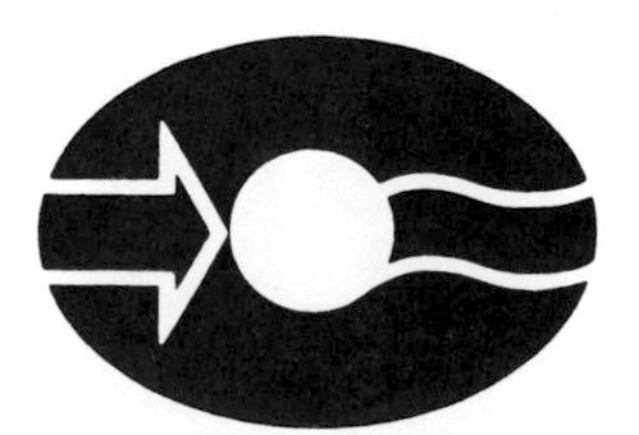

142

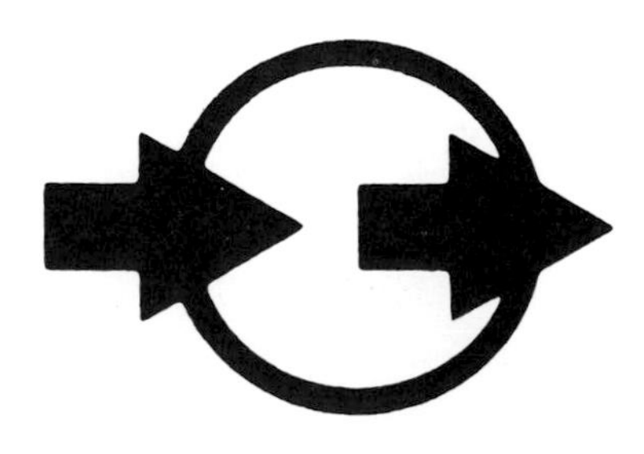

143

144	**Metagraphics Software Corporation**	Mountain View, California *Computer graphics software* Design: Jack Davis
145	**Data Memory Incorporated**	Santa Clara, California *Video disc systems* Design: Gruyé Associates
146	**Daytronic Corporation**	Miamisburg, Ohio *Industrial data acquisition & control systems*
147	**Evaluation and Planning Systems**	Windham, New Hampshire *Decision support software*
148	**Nippon Kogaku K.K.**	Tokyo, Japan *Nikon camera auto-focus* Design: Yusaku Kamekura
149	**Nippon Kogaku, K.K.**	Tokyo, Japan *Nikon camera auto-focus* Design: Yusaku Kamekura

144

145

146

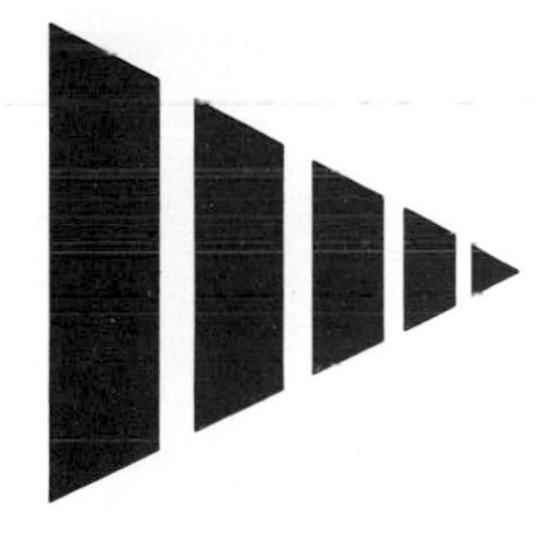

147

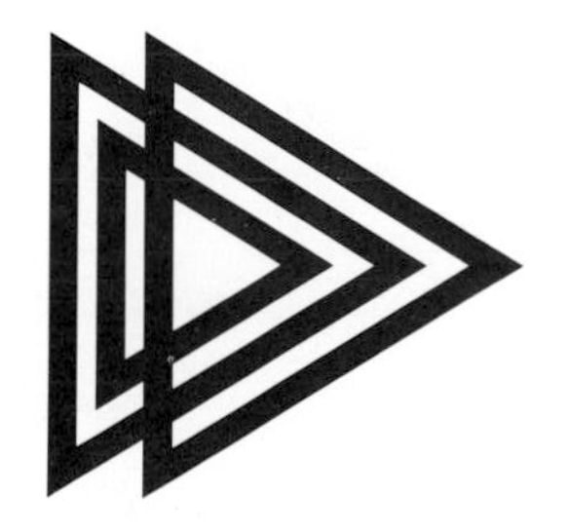

148

149

 Particle Measuring Systems

Boulder, Colorado
Laser particle-size spectrometers
Design: Nancy Towne

151	**Micronic Corporation**	Los Gatos, California *X-Ray lithography systems* Design: Norman Orr
152	**Cuesta Systems**	San Luis Obispo, California *Power Protection equipment for microcomputers* Design: Dave Wells
153	**Mountain Computer**	Scotts Valley, California *Computer peripherals* Design: Lutat, Battey & Associates
154	**Tritec Industries**	Sunnyvale, California *Chemical dispense pumps* Design: Karen Holman
155	**Timberline Systems**	Portland, Oregon *Business software* Design: Metamorphic Design
156	**Skyland Systems**	Scotts Valley, California *Power controllers for microcomputers* Design: Lynn Marsh

151

152

153

154

155

156

157	**Expert Systems**	Redmond, Washington *Business & educational software*
158	**Digital Pathways**	Palo Alto, California *Voice communication software for PCs* Design: Rodney's
159	**Johnson Associates Software**	Redding, California *Health care software* Design: Geodart Advertising
160	**XYCOM Incorporated**	Saline, Michigan *Harsh environment electronic equipment*
161	**Hawke Electronics**	Middlesex, England *Electronic component distributors* Design: Thorn Advertising
162	**Damomics**	Corning, New York *Microcomputer software* Design: Ronald A. Friedman

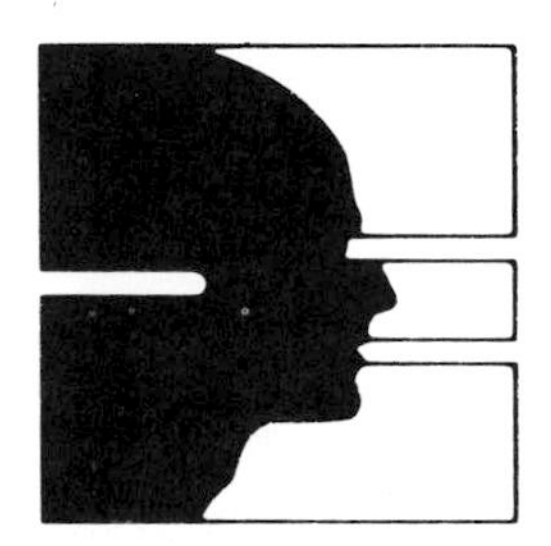

157

158

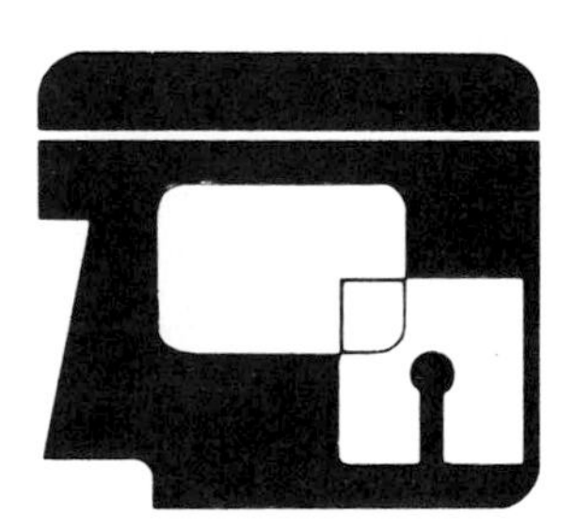

159

160

161

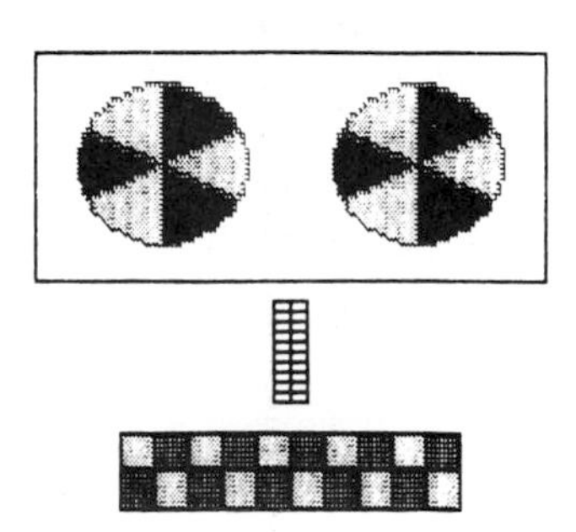

162

163 **Morrow**

San Leandro, California
Microcomputer systems & peripherals
Design: Michael Mabry

164	**Alcor Systems**	Garland, Texas *Computer program systems*
165	**Kapstrom Incorporated**	Dallas, Texas *Educational software* Design: Ted J. Karch
166	**Soft Solutions**	San Francisco, California *Computer software*
167	**MC/N Incorporated**	San Rafael, California *Floppy disk jewelry* Design: Christopher Van Dyke, Mary Shepard
168	**Software Distributors Clearinghouse**	Georgetown, Texas *Public domain software distributor* Design: Marketing Edge
169	**IMSI**	San Rafael, California *Software publisher* Design: Naganuma Design

164

165

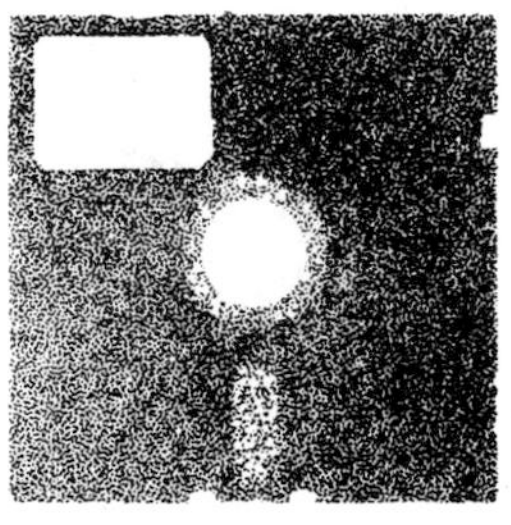

166

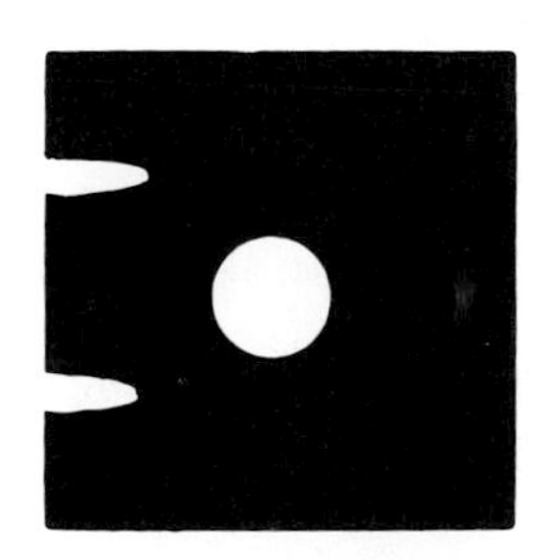

167

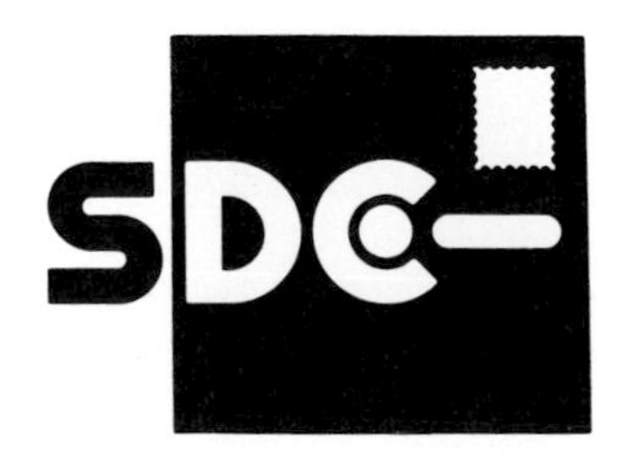

168

169

170	**Ann Arbor Software**	Ann Arbor, Michigan *Business software for the IBM PC* Design: In-house
171	**National Software Institute**	Rancho Bernardo, California *Computer software development* Design: Jay Hanson Design Associates
172	**American Training International**	Manhattan Beach, California *Computer training* Design: Jay Hanson Design Associates
173	**Floppy Disk Services**	Lawrenceville, New Jersey *Floppy & hard disk subsystems* Design: Airon Advertising
174	**Shaffstall Corporation**	Indianapolis, Indiana *Computer peripherals* Design: Art Craft Press
175	**FairCom**	Columbia, Missouri *Computer software development* Design: John Lynch

170

171

172

173

174

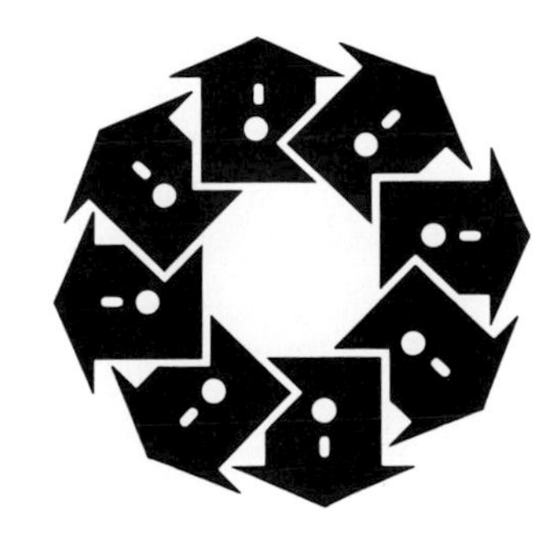

175

 SmithKline Bioscience Laboratory

Van Nuys, California
National network of clinical reference labs
Design: Primo Angeli Graphics

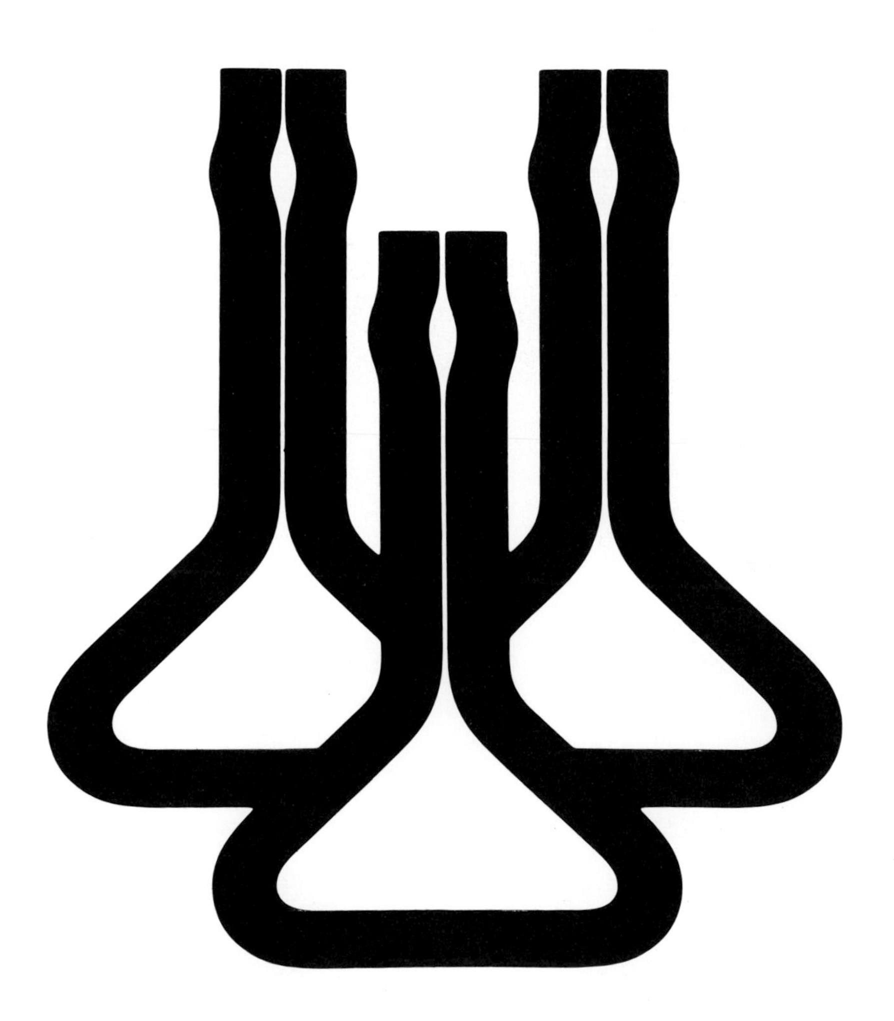

177	**Omega Software**	Round Rock, Texas *Software for the IBM PC* Design: Robert A. Miller
178	**ICR FutureSoft**	Orange Park, Florida *Program generators*
179	**Key Solutions**	Campbell, California *Software for the IBM PC*
180	**PC Centre**	Sunnyvale, California *Personal computer sales and consulting*
181	**Personal Computer Training Programs**	Los Angeles, California *Computer training* Design: J. Michael Hook Design
182	**The Purchasing Agent**	Mountain View, California *Microcomputer buying agent* Design: Perry-Kane

177

178

179

180

181

182

183	**National Instruments**	Austin, Texas *Automated instrumentation interfaces* Design: McLane Advertising
184	**Palantir Software**	Houston, Texas *Integrated office automation software*
185	**American Magnetics Corporation**	Carson, California *Digital magnetic tape heads*
186	**Amerisoft**	Petaluma, California *Dataprocessors* Design: In-house
187	**Recruit Company**	Tokyo, Japan *Data processing* Design: Yusaku Kamekura
188	**Condor Computer Corporation**	Palo Alto, California *Data management software*

183

184

185

186

187

188

189 **Lockheed-California / Skunkworks**

Burbank, California
Aircraft design & operations research
Design: In-house

190	**Meadowlark Optics**	Longmont, Colorado *Polarization optics for laser systems* Design: Howard Crosslen
191	**Avocet Systems**	Dover, Delaware *Software development tools for industry*
192	**Wolfdata**	Chelmsford, Massachusetts *Modems*
193	**Oryx Systems**	Wausau, Wisconsin *Microcomputer hardware/software sales* Design: Talal A. Kheir
194	**Lionheart**	Alburg, Vermont *Computer software publisher* Design: Richard Stevenson
195	**Frontier Technologies Corporation**	Milwaukee, Wisconsin *Graphic controllers for the IBM PC* Design: In-house

190

191

192

193

194

195

196	**MacLion / Computer Software Design**	Anaheim, California *Relational data base manager system software* Design: Chuck Bennett
197	**Novation Incorporated**	Chatsworth, California *Modems* Design: Gene Dieckhoner
198	**Mouse Systems Corporation**	Santa Clara, California *PC mouse*
199	**Mouse House Incorporated**	Berkeley, California *Manufacturer of computer mice (input device)* Design: In-house
200	**Business Solutions Incorporated**	Kings Park, New York *Computer software*
201	**Vexilar**	New York, New York *Sonar equipment* Design: Yusaku Kamekura

196

197

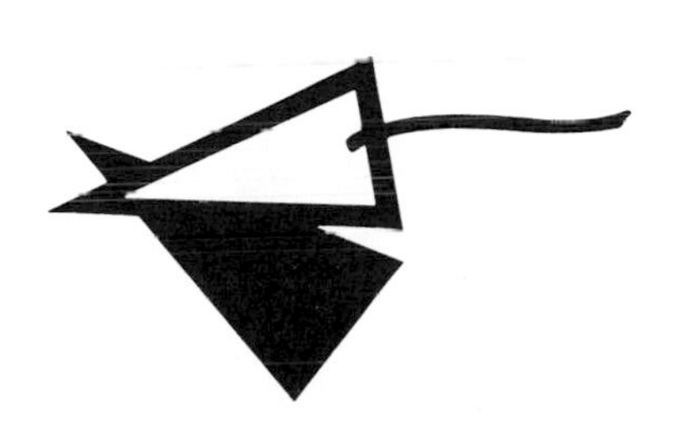

198

199

200

201

202	**Corona Data Systems**	Thousand Oaks, California *IMB PC-compatible portable computers*
203	**Solarex Corporation**	Rockville, Maryland *Photovoltaic systems*
204	**Rambow Enterprises**	Anchorage, Alaska *Business software* Design: In-house
205	**Texas Instruments**	Dallas, Texas *Electronic equipment*
206	**General Photonics Corporation**	Santa Clara, California *Lasers*
207	**PC Software**	San Diego, California *Computer software* Design: Joseph Juhasz, Keith Oswald

202

203

204

205

206

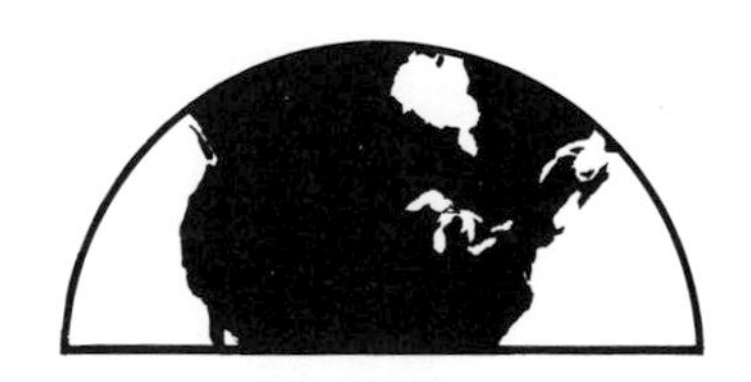

207

208	**Management Science America**	Atlanta, Georgia *Business & educational software* Design: In-house
209	**Quinsept**	Lexington, Massachusetts *Genealogy research software*
210	**RealSoft Systems**	Aspen, Colorado *Real estate software* Design: Get Graphics
211	**Lifetree Software**	Monterey, California *Word processors* Design: RJ Wecker Graphics
212	**Barr Systems**	Raleigh, North Carolina *Communications hardware/software* Design: Rasor & Rasor
213	**Island Graphics**	Honolulu, Hawaii *Computer graphics software* Design: Rand Schulman

208

209

210

211

212

213

214	**RoseSoft Incorporated**	Seattle, Washington *Computer software* Design: Ben Dennis & Associates
215	**Sweet·P / Enter Computer Company**	San Diego, California *Computer software* Design: In-house
216	**Metamorphics**	Bala Cynwyd, Pennsylvania *Computer software* Design: Geri Zhiss
217	**Whitesmiths Ltd.**	Concord, Massachusetts *Computer software development*
218	**Mupac Corporation**	Brockton, Massachusetts *Integrated circuits*
219	**Friends Software Corporation**	Berkeley, California *Computer software* Design: Jonathan Nix

214

215

216

217

218

219

220 **Silvar-Lisco**
Menlo Park, California
Computer-aided engineering software

221 **XCOMP Incorporated**
San Diego, California
Local area network for IBM PCs
Design: Busch & Associates

222 **Post Technologies**
Menlo Park, California
Electronic mail terminals
Design: David Kelly Design

223 **Hongkong Bank**
Hong Kong
Automatic teller machine network
Design: Henry Steiner

224 **Cenna Technology**
Sandy, Utah
Floppy disks
Design: Barbara Rasekhi

225 **Tallgrass Technologies Corporation**
Overland Park, Kansas
Microcomputer backup subsystems
Design: In-house

220

221

222

223

224

225

 Sym Pro

Albany, California
Business computer systems
Design: Jerry Kuyper Design

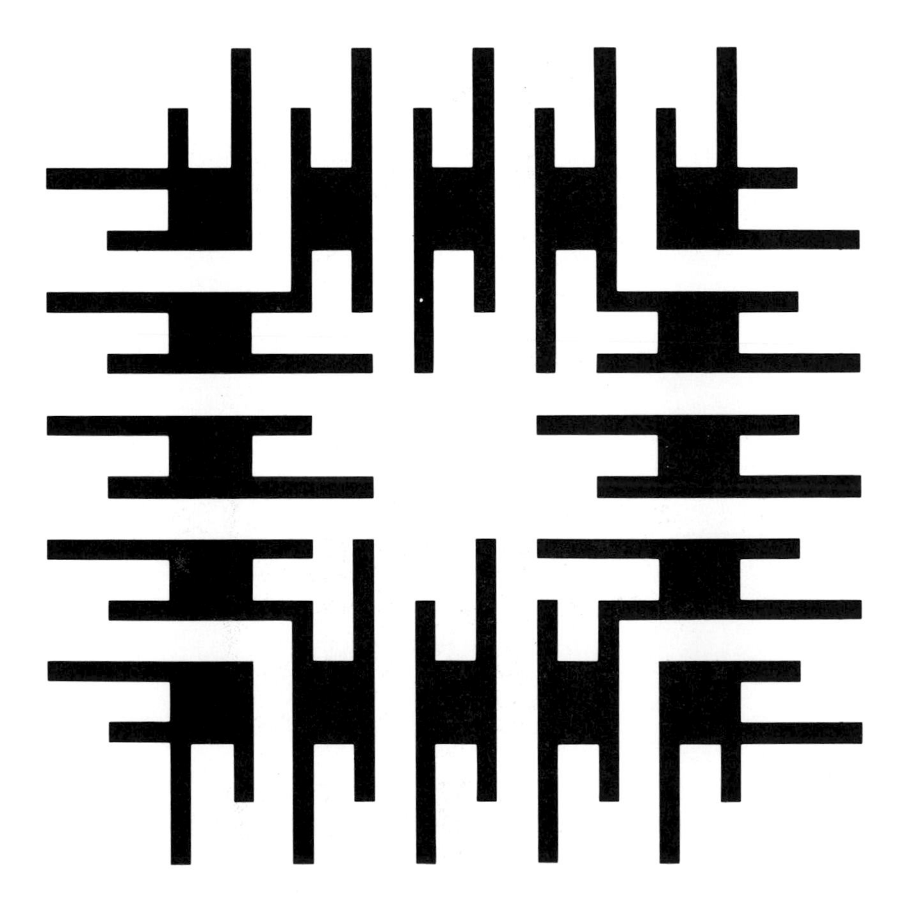

227	**BOC Health Care**	New York, New York *Life support systems & intravascular components* Design: Anspach Grossman Portugal
228	**FORTH Incorporated**	Hermosa Beach, California *Electronic equipment* Design: James Cross
229	**Balazs Analytical Laboratory**	Mountain View, California *Instrumental & chemical analysis* Design: Marjorie K. Balazs
230	**Logical Software**	Cambridge, Massachusetts *UNIX user interface software* Design: Gill Fishman Associates
231	**On-Line Systems**	San Francisco, California *Interactive data processing services* Design: Logan, Carey & Rehag
232	**National Memory Systems**	Livermore, California *Computer hardware*

227

228

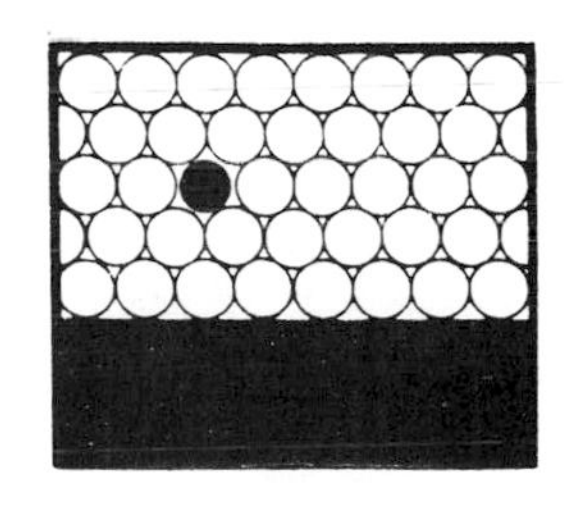

229

230

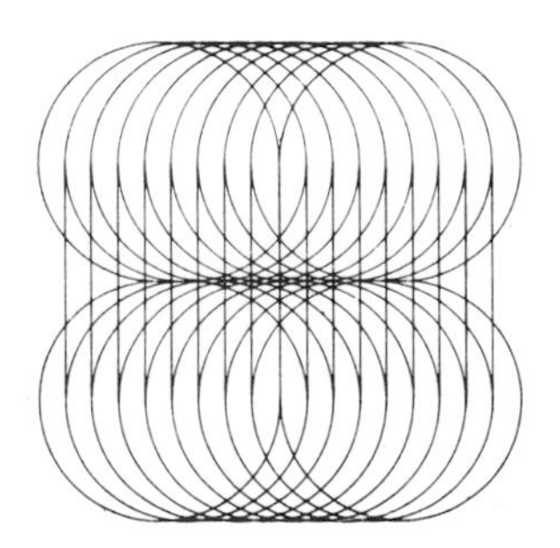

231

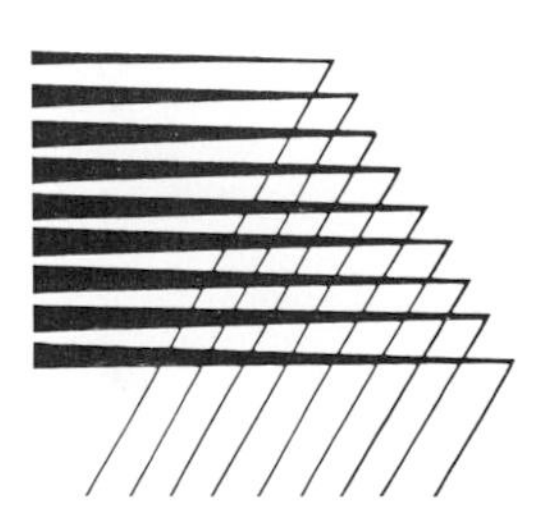

232

233	**Microdex Corporation**	Tucson, Arizona *Computer-assisted design systems* Design: CB Intersol
234	**Lee Payne Associates**	Atlanta, Georgia *Industrial design for technology companies* Design: Lee Payne Associates
235	**Priam**	San Jose, California *Winchester disk drives* Design: Primo Angeli
236	**Cleo Software**	Rockford, Illinois *Software for micro-to-mainframe connection*
237	**National Semiconductor Corporation**	Santa Clara, California *32 bit microprocessor* Design: Gordon Mortensen
238	**Genus Incorporated**	Mountain View, California *Integrated circuit fabrication equipment* Design: Owens/Lutter Advertising

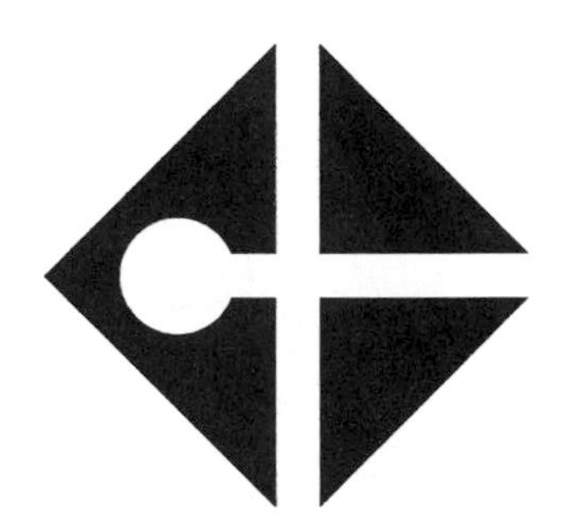

233

234

235

236

237

238

239	**Davidson & Associates**	Rancho Palos Verdes, California *Educational software publisher* Design: Dali Bahat
240	**Nanodata Computer Corporation**	Buffalo, New York *Mainframe digital computers* Design: In-house
241	**Voltrax**	Troy, Michigan *Text-to-speech synthesizers*
242	**Winegard Company**	Burlington, Iowa *Television reception products*
243	**Finnigan Corporation**	San Jose, California *Mass spectrometers* Design: In-house
244	**Laser Precision Corporation**	Utica, New York *Laser power/energy meters* Design: Graystone Designs

239

240

241

242

243

244

245 **Gifford Computer Systems**

San Leandro, California
Operating systems software
Design: Lawrence Bender & Associates

246	**Pacific Scientific**	Anaheim, California *Electronic instruments* Design: Bud Willis
247	**Visidata Corporation**	Fremont, California *Computer terminals* Design: Gruyé Associates
248	**Triad Systems**	Sunnyvale, California *Computer hardware/software* Design: Bill Stevens, Don Ruder, Hank Gay
249	**Conographic Corporation**	Irvine, California *Computer graphics hardware/software* Design: Luis Villalobos
250	**Datapoint Corporation**	San Antonio, Texas *Office automation systems* Design: In-house
251	**Network Research Corporation**	Los Angeles, California *Local area network software* Design: Sarah Pugh

246

247

248

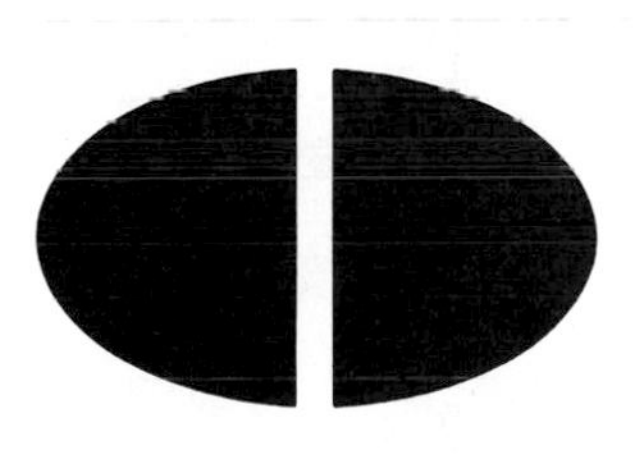

249

250

251

252	**Synactics Corporation**	Santa Clara, California *Office automation software* Design: Chapman - Sabatella Advertising
253	**Tonko**	Campbell, California *Process machinery* Design: Lawrence Bender & Associates
254	**Xinix Incorporated**	Santa Clara, California *Instruments for semiconductor process control* Design: George Opperman
255	**The Hilliard Corporation**	Elmira, New York *Oil filtration & reclamation systems* Design: Grphiden, Inc.
256	**Microtec**	Irvine, California *Software development*
257	**Advanced Ideas**	Berkeley, California *Educational software* Design: Parallel Communications

252

253

254

255

256

257

258 **AutoCAD**

Mill Valley, California
CAD/CAM/CAE graphics software
Design: John Buckley
(drawn on a TI portable personal computer using AutoCAD software and printed with a HP 7475 plotter)

259	**Starsoft**	Mountain View, California *Software distributors* Design: Jerry Blank
260	**Microsoft Corporation**	Bellevue, Washington *Software for microcomputers* Design: David Strong Design Group
261	**Pearlsoft**	Wilsonville, Oregon *Computer software* Design: April Greiman

259

MICROSOFT

260

pearlsoft

261

262 **Falcon Safety Products**
New York, New York
Portable compressed-gas dust removal devices

263 **Eagle Computer**
Los Gatos, California
IBM PC compatible and networking products
Design: Paul Sinn

264 **Rolm Corporation**
Santa Clara, California
Logo for telecommunications newsletter
Design: Shelton Fong

202

263

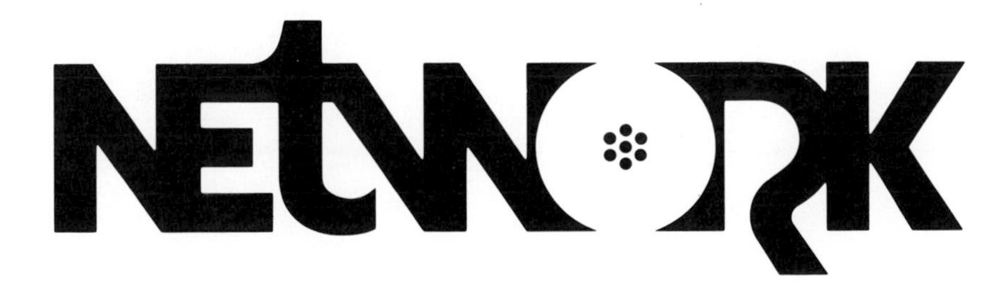

264

265 **Telegenex Corporation**

Atlanta, Georgia
Bit error rate testing equipment
Design: David Laufer Associates

266	**USENIX**	El Cerrito, California *UNIX users' group* Design: Aaron Marcus & Associates
267	**Metheus Corporation**	Hillsboro, Oregon *Computer-aided engineering workstations* Design: Steve Wittenbrock
268	**Continental Telecom**	Atlanta, Georgia *Integrated telecommunications systems* Design: Anspach Grossman Portugal

266

METHEUS

267

CONTEL

268

269	**Audiotronics**	Memphis, Tennessee *Audio mixing consoles*
270	**The Charles Stark Draper Laboratory**	Cambridge, Massachusetts *Guidance and navigation control system R&D*
271	**Laser Engineering Incorporated**	Waltham, Massachusetts *Carbon dioxide lasers* Design: G.J. Khoury, Inc.
272	**Coherent**	Palo Alto, California *Lasers and laser systems* Design: Dean Smith
273	**CW Communications**	Framingham, Massachusetts *Computer magazines & newspapers*
274	**Carsonic / Division of Fujitsu Ten**	Torrance, California *Audio equipment* Design: Huerta Design Associates

269

270

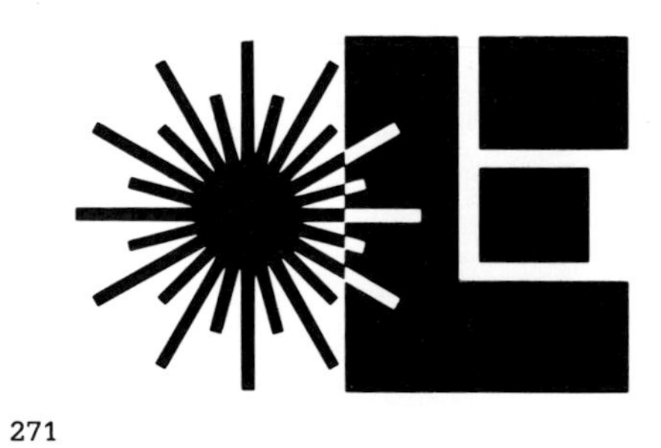

271

272

273

274

275	**Hayes Microcomputer Products**	Norcross, Georgia *Hardware/software for microcomputers* Design: Lee Payne, Jack Snoke
276	**Genicso Technology**	Cypress, California *Computer graphics & imaging products*
277	**Braemar Computer Corporation**	Burnsville, Minnesota *Computers* Design: Sheila Chin
278	**Zyvex Corporation**	San Jose, California *Sub-contract assembly & manufacturing* Design: Pam Will
279	**National Applied Computer Technology**	Salt Lake City, Utah *Computer hardware* Design: Steven R. Grigg
280	**Dibble Electronics**	San Diego, California *Multi-layer printed circuit boards* Design: Jon Scharnborg & Associates

275

276

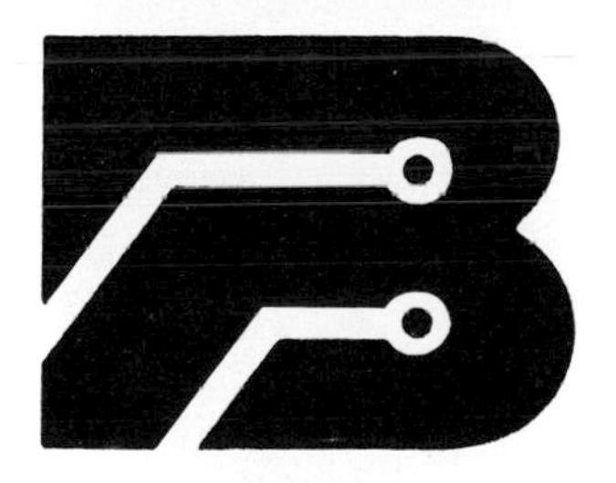

277

278

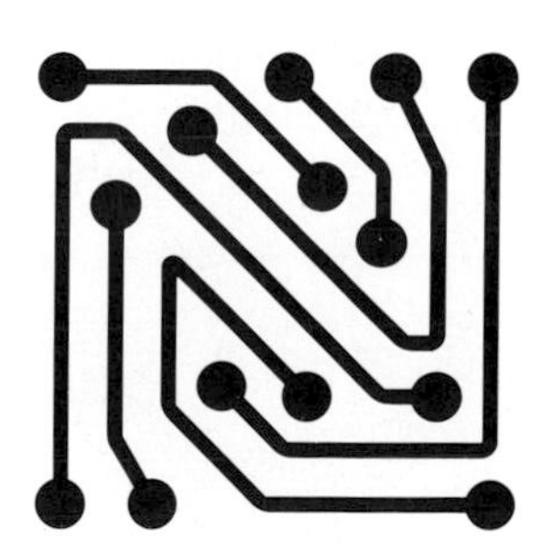

279

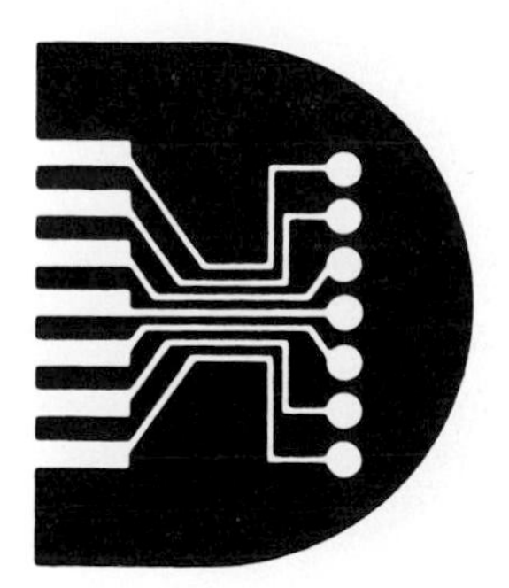

280

281 **CADAM Inc**
Burbank, California
Computer-assisted design software
Design: In-house

282 **Hongkong Telephone Company**
Hong Kong
Communication services
Design: Henry Steiner

283 **Mikros Systems Corporation**
Mercerville, New Jersey
Military/areospace microcomputers

284 **Treffers Precision**
Phoenix, Arizona
Precision machining & shafting equipment
Design: John Treffers

285 **SYSCON Corporation**
Washington, D.C.
Systems engineering computer systems

286 **Unitronix Corporation**
Sparta, New Jersey
Microcomputer products
Design: Advertising/Marketing Associates

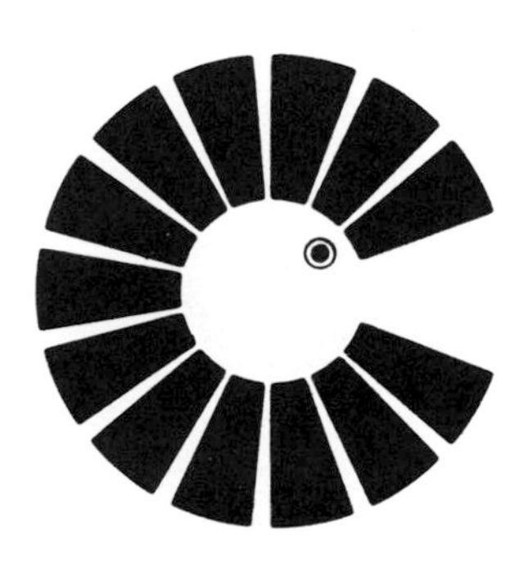
281

282

283

284

285

286

287	**Reston Computer Group**	Reston, Virginia *Computer books & software* Design: Diane Hoppman
288	**Vector Electronic Company**	Sylmar, California *Printed circuit boards & data storage systems*
289	**Advanced Control Components**	Clinton, New Jresey *Semiconductor diode and microwave components*
290	**The Aerospace Corporation**	Los Angeles, California *Engineer services for military space systems*
291	**Zeta Laboratories**	Santa Clara, California *Microwave and R.F. components* Design: Dewayne Cottrill
292	**Uniphase Corporation**	Sunnyvale, California *Helium-neon lasers & laser systems* Design: Watson Southwest

287

288

289

290

291

292

293 **Nova Graphics International Corporation** Austin, Texas
Computer graphics software
Design: Larry Smitherman

294 **Norango Computer Systems**
Toronto, Ontario, Canada
Business software
Design: Spencer/Francey Incorporated

295 **ExperTelligence**
Santa Barbara, California
Artificial intelligence langauges
Design: Tom Adler Design

296 **Profit Plus Software**
Sausalito, California
Computer software
Design: Crawshaw & Company Design

297 **Perception Electronics**
San Jose, California
Electronic equipment
Design: Patrick Mountain Design

298 **Catalyst Technologies**
Santa Clara, California
Business support systems
Design: Gerald Reis & Company

299 **Stanford Business Software**
Palo Alto, California
Software development
Design: Sharrie Brooks

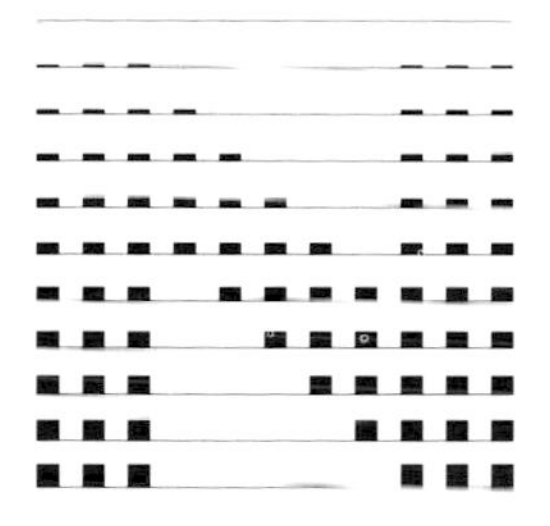

294

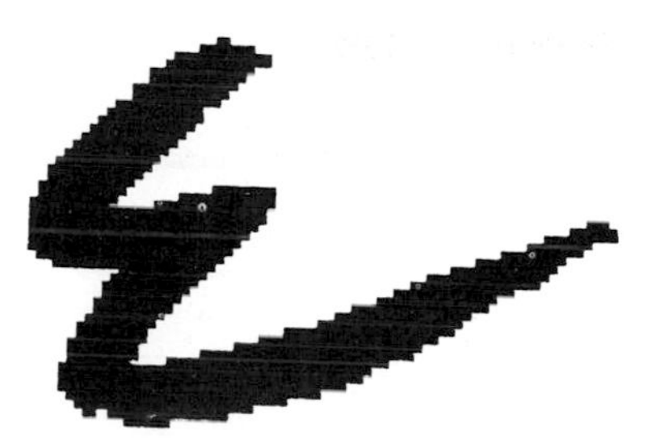

295

296

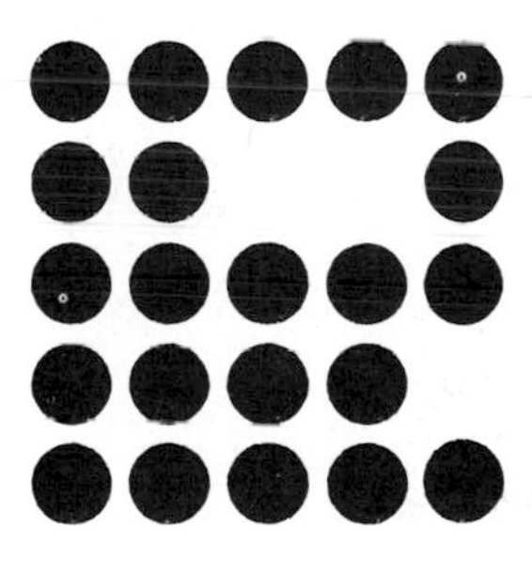

297

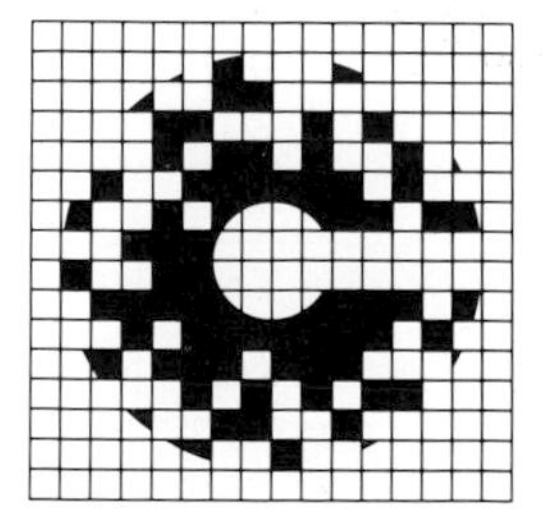

298

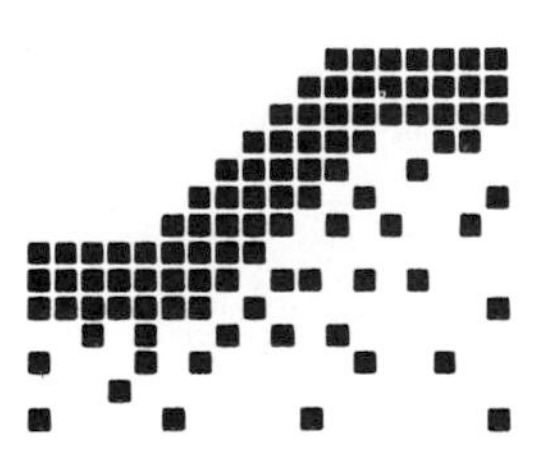

299

300	**Computer Systems Corporation**	Indianapolis, Indiana *Computer software, terminals & printers* Design: Jim Brooks
301	**Matthews Computer Corporation**	San Diego, California *Computer software & systems* Design: Gruyé Associates
302	**Versatec**	Santa Clara, California *Electrostatic printers & plotters* Design: Anthony P.H. Chan
303	**Finell Systems**	San Jose, California *Servo drive & control products*
304	**Software Systems**	Provo, Utah *Computer software for accounting* Design: Steven R. Grigg
305	**Seequa Computer Corporation**	Odenton, Maryland *IBM compatible computers*

300

301

302

303

304

305

306	**Lawrence Livermore National Laboratory**	Livermore, California *Nuclear weapon & fusion energy research* Design: In-house
307	**National Semiconductor**	Santa Clara, California *Semiconductors* Design: Ken Parkhurst
308	**Varian Associates**	Palo Alto, California *Computer & electronics equipment* Design: Larry Klein
309	**MITEQ**	Hauppauge, New York *Telecommunications equipment*
310	**Quantum Software Systems**	Ottawa, Ontario, Canada *Computer software*
311	**Argonne National Laboratory**	Argonne, Illinois *Multidisciplinary research center* Design: In-house

306

307

308

309

310

311

312	**Photographic Sciences Corporation**	Webster, New York *Color graphic recording cameras* Design: Ice Communications
313	**Quadratron Systems**	Encino, California *Office automation software* Design: Norm Abbey
314	**Brooks Automation**	North Billercia, Massachusetts *Automated wafer transport systems* Design: Bob Kelly
315	**Digigraphic Systems Corporation**	Minnetonka, Minnesota *Computer-based management information systems*
316	**Arrays Incorporated**	Los Angeles, California *Computer software* Design: Ellis, Ross, Tani Advertising
317	**Indus Systems**	Chatsworth, California *Computer peripherals* Design: Huerta Design Associates

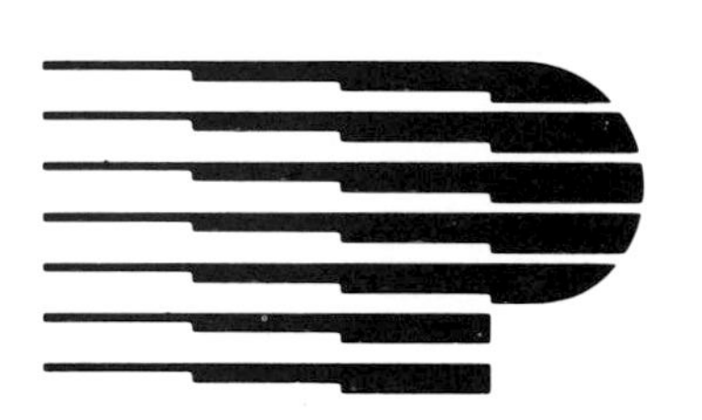

312

313

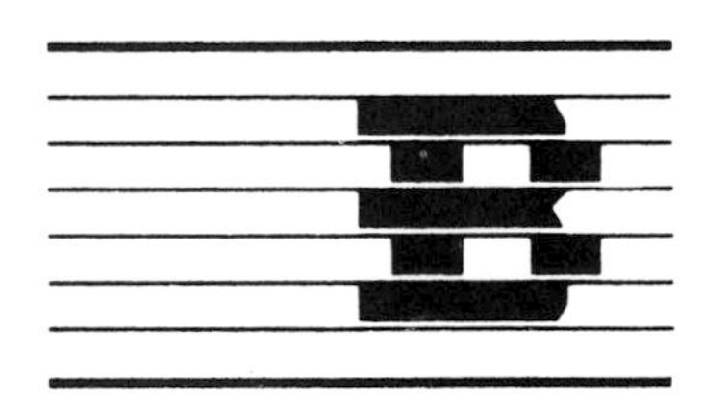

314

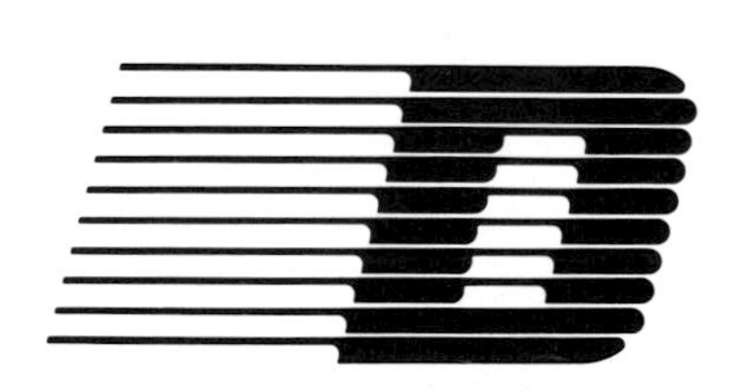

315

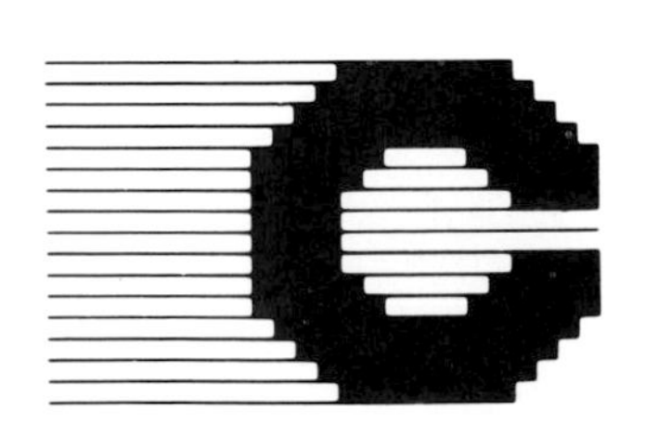

316

317

318	**Novell**	Orem, Utah *Local area network hardware/software* Design: Tandem Studios
319	**SIGGRAPH**	Chicago, Illinois *Computer graphics association*
320	**Queue Incorporated**	Bridgeport, Connecticut *Educational software* Design: Tom Richardson Associates
321	**Alpha Software Corporation**	Burlington, Massachusetts *Integrated productivity software*
322	**CompuSoft Publishing**	El Cajon, California *Computer books*
323	**Conrac Division of Conrac Corporation**	Covina, California *Video monitors for computer terminals* Design: Harry Murphy & Friends

318

319

320

321

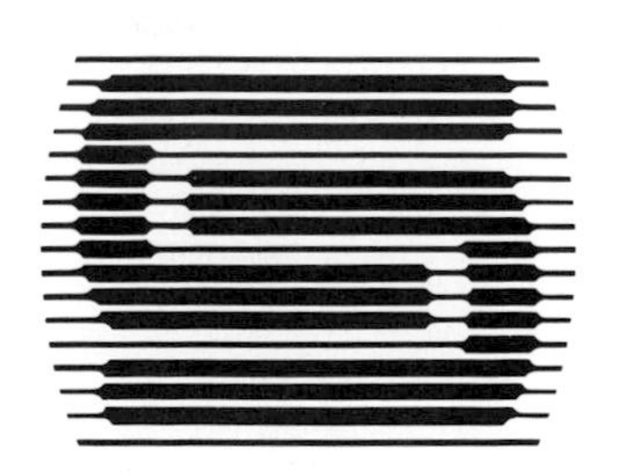

322

323

324 **Texwipe**
Upper Saddle River, New Jersey
Precision cleaning products & wipers
Design: Cook and Shanosky

325 **Technical Instrument Company**
San Francisco, California
Microscopes & precision measuring systems
Design: Jorgensen Design

326 **Software Express**
Houston, Texas
Software sales
Design: Chris Hill

327 **Standard Microsystems**
Hauppauge, New York
MOS integrated circuits
Design: Lawrence Bender & Associates

328 **Scope Incorporated**
Reston, Virginia
Digital communication

329 **Zebec Data Corporation**
Houston, Texas
Computer timesharing
Design: Chris Hill

324

325

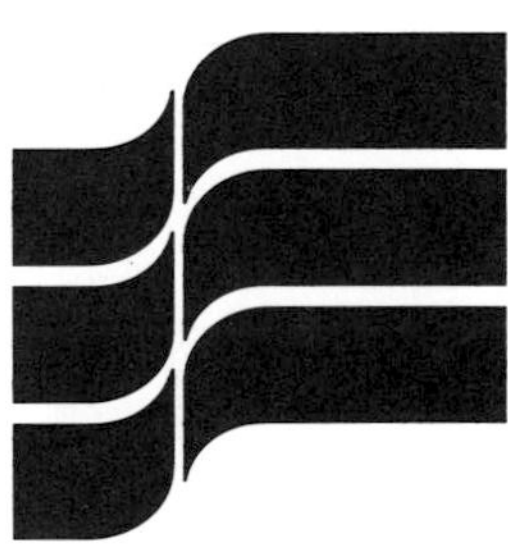

326

327

328

329

330 **Jabend Corporation**

Seattle, Washington
Computer software
Design: Lawrence Bender & Associates

331	**Applied Image**	Rochester, New York *Precision electronic and optical equipment* Design: Bruno B. Glavich
332	**AG Associates**	Palo Alto, California *Rapid thermal processing systems*
333	**George Lithograph**	San Francisco, California *Design & printing for high tech corporations* Design: George Collopy
333	**Triemco**	France *Telecommunications systems* Design: Gruyé Associates
335	**Mizar Incorporated**	St. Paul, Minnesota *Software development systems*
336	**VLSI Technology**	San Jose, California *Application specific integrated circuits* Design: Sally DeSpain

331

332

333

334

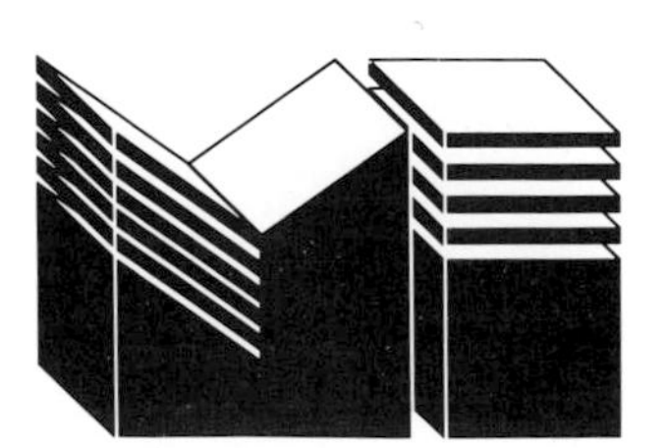

335

336

337 **Information NOW**
Malibu, California
Marketing consultation via computer network
Design: Jerry Jankowski

338 **Blue Chip Software**
Canoga Park, California
Financial market simulation software
Design: George Hand

339 **Picture Element Limited**
Palo Alto, California
Video sequence processors
Design: Lawrence Bender & Associates

340 **GVO Incorporated**
Palo Alto, California
Design & engineering consulting
Design: Gruye Associates

341 **VIA Systems**
North Billercia, Massachusetts
Turnkey CAD/CAM/CAE computer systems

342 **LaserLink / Landart Systems**
New York, New York
Laser printer devices
Design: Aaron Marcus & Associates

337

338

339

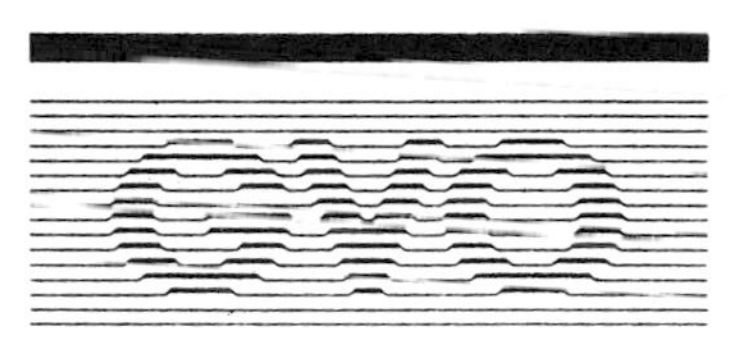

340

341

342

343	**STB Systems**	Richardson, Texas *IBM PC peripherals* Design: Keeton & Rich
344	**Advanced Electronic Design**	Sunnyvale, California *Computer workstations and disk controllers* Design: Reiser Williams DeYong
345	**Logical Data Management**	Covina, California *Software for data management* Design: Mark Wood
346	**Data Management Labs**	San Jose, California *Winchester disk subsystems*
347	**Dynamic Microprocessor Associates**	New York, New York *Business software* Design: Jerry Wilke Design
348	**International Microelectronic Products**	San Jose, California *Custom integrated circuits* Design: Lawrence Bender & Associates

343

344

345

346

347

348

349	**Fisher Business Systems**	Atlanta, Georgia *Computer software*
350	**TRW Incorporated**	Cleveland, Ohio *Electronic equipment* Design: Siegal & Gale
351	**Automatic Data Processing**	Roseland, New Jersey *Computing & information services* Design: Arthur Boden
352	**DTC**	Campbell, California *Data terminals & communications* Design: David Brandenburg
353	**Computer Aided Engineering**	Sunnyvale, California *Workstations for integrated circuit design* Design: Lawrence Bender & Associates
354	**NCA Corporation**	Santa Clara, California *Computer timesharing services* Design: Lawrence Bender & Associates

349

350

351

352

353

354

355 **Bytec-Comterm**
Pointe Claire, Canada
Data communications & computer terminals

---

356 **Robin**
Atlanta, Georgia
Process management systems
Design: Gruyé Associates

---

357 **Assemco**
Palo Alto, California
High technology metal
Design: A.R. Stein, Dennis Moore

355

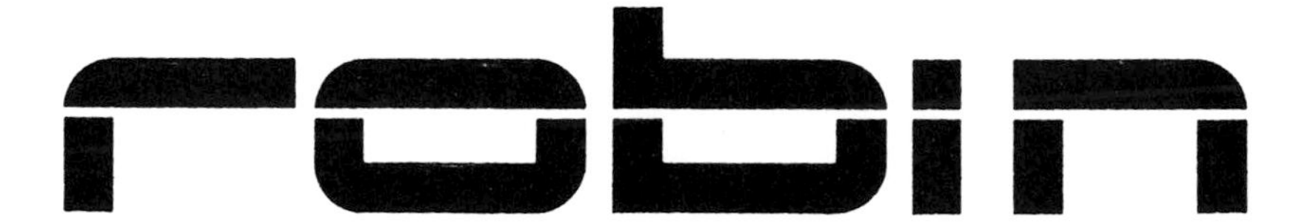

356

357

358	**Stearns Computer Systems**	Minneapolis, Minnesota *Microcomputers* Design: Nancy Rice
359	**Micromax Systems**	San Diego, California *Printed circuit add-on cards*
360	**Smart Cable / IQ Technologies**	Bellevue, Washington *Interface cables* Design: Nagel Design

358

359

360

361	**Control Data Corporation**	Minneapolis, Minnesota *Computers*
362	**Computer Dynamics**	Greer, South Carolina *Computer-based electronics* Design: Cathy Adams Rodby
363	**Adams-Russell**	Burlington, Massachusetts *RF & microwave signal processing components* Design: Seymour Levy
364	**Seagate**	Scotts Valley, California *Disk drives*
365	**Integrated Technologies**	Acushnet, Massachusetts *Electronic lithography systems* Design: Hendrik F. Bok
366	**Technicom Systems**	Whippany, New Jersey *Communications equipment*

361

362

363

364

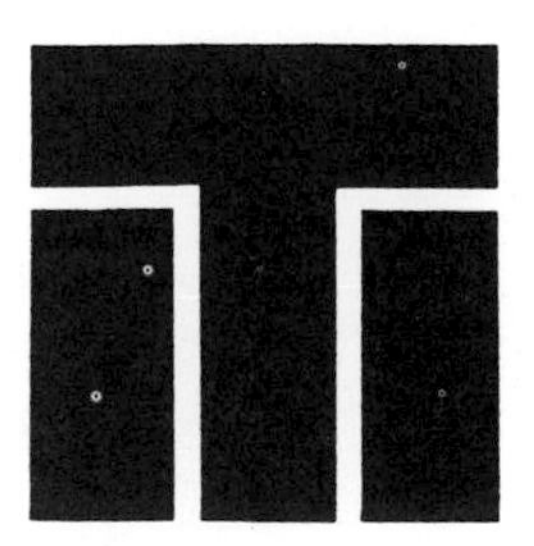

365

366

367	**Datablend**	Woodinville, Washington *Industrial software*
368	**Precision Data Products**	Grand Rapids, Michigan *Information processing products*
369	**MARC Software International**	Palo Alto, California *Software research & development*
370	**Mitel Corporation**	Ontario, Canada *CMOS integrated circuits*
371	**Ellis Computing**	San Francisco, California *Microcomputer software* Design: Bonfield & Associates
372	**Dynachem Corporation**	Tustin, California *Chemicals for printed circuit manufacture*

367

368

369

370

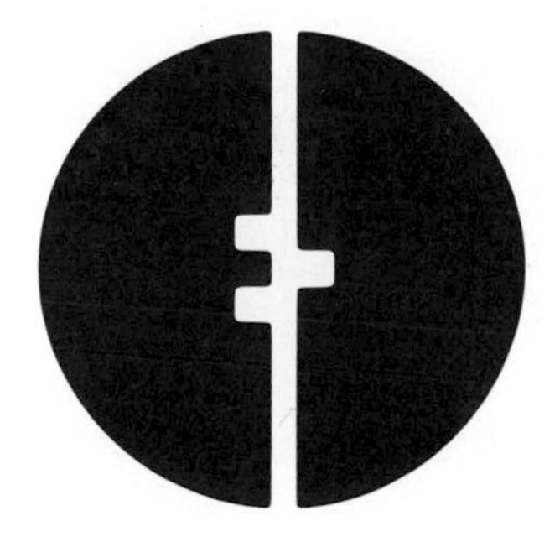

371

372

373	**Kadak Products Ltd.**	Vancouver, Canada *Real-time software engineering* Design: W.L. Renwick
374	**XonTech Incorporated**	Van Nuys, California *Meterological data acquisition equipment*
375	**Personal Systems Technology**	Irvine, California *Computer hardware/software* Design: Bobbi Gentry
376	**Electronics Materials Packaging**	Chanhassen, Minnesota *Semiconductor industry labware, racks & trays* Design: Coulter Advertising
377	**Gaertner Scientific Corporation**	Chicago, Illinois *Electro-optical instrumentation*
378	**Atlantic Research Corporation**	Alexandria, Virginia *Solid rocket propulsion & electronic systems* Design: In-house

373

374

375

376

377

378

379 **Pure Data Limited**

Markham, Ontario, Canada
Expansion boards for the IBM PC
Design: Jaitee Communication Art

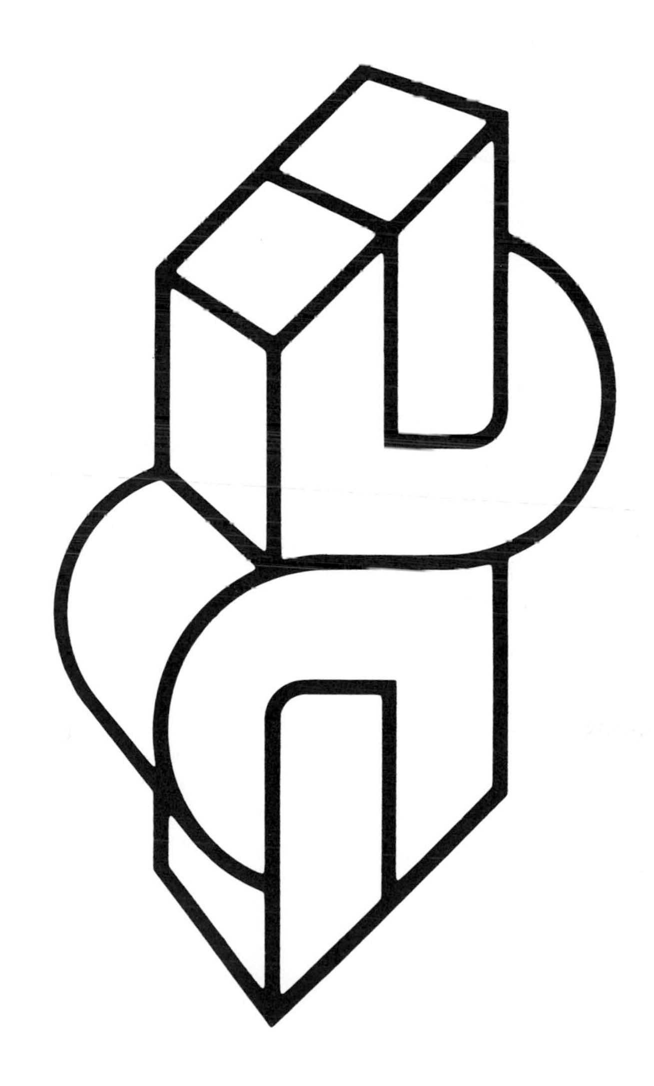

380	**Exar Corporation**	Sunnyvale, California *Custom integrated circuits* Design: Yashi Okita Design
381	**Xyvision**	Woburn, Massachusetts *Electronic composition and pagination systems*
382	**Logical Business Machines**	Sunnyvale, California *Business computers* Design: Russell Leong Design
383	**Lehr/Fiegehen Technologies**	Carson City, Nevada *Computer products for business* Design: Rich Lebedeff
384	**Macola Incorporated**	Marion, Ohio *Computer software* Design: In-house
385	**Modular Instruments Incorporated**	Southeastern, Pennsylvania *Modular interface for the IBM PC* Design: Ann Mullikin

380

381

382

383

384

385

386	**State of the Art**	Costa Mesa, California *Accounting software for microcomputers* Design: Design Center
387	**Electro Rent Corporation**	Burbank, California *Electronic equipment distributor* Design: Gaby Moussa
388	**Chicago Laser Systems**	Chicago, Illinois *Laser trim systems* Design: Joseph Milani
389	**Interactive Structures**	Bala Cynwyd, Pennsylvania *Computer printer interfaces* Design: In-house
390	**Software Research Corporation**	Victoria, B.C., Canada *Educational hardware/software* Design: Silk Questo
391	**U.S. Robotics**	Chicago, Illinois *Modems* Design: Beda-Ross Associates

386

387

388

389

390

391

392	**Lexisoft**	Davis, California *Word processing software* Design: David Saake
393	**National Semiconductor**	Santa Clara, California *8 bit microprocessor* Design: Gordon Mortensen
394	**Terminal Data Corporation**	Rockville, Maryland *Dataprocessing equipment*
395	**Alpha Graphix**	Los Angeles, California *Computerized typography* Design: Don Weller
396	**Chalk Board Incorporated**	Atlanta, Georgia *Computer software* Design: Taylor & Taylor
397	**Quality Software**	Chatsworth, California *Computer software publishers* Design: Les Foster

392

393

394

395

396

397

398	**Integrated Solutions**	Campbell, California *Minicomputers* Design: David Kelley Design
399	**Shipley Company**	Newton, Massachusetts *Photo resist for printed circuit board fabrication* Design: Ingalls Associates
400	**Interactive Development Environments**	San Francisco, California *Software development* Design: Peter Rocha
401	**ADE Corporation**	Newton, Massachusetts *Instrumentation & test equipment* Design: Salame Associates
402	**Context Management Systems**	Torrance, California *Integrated software for personal computers* Design: Michael Cohen, Michael Ofstein
403	**TECH-S Incorporated**	Livonia, Michigan *Data acquisition peripherals* Design: Dwayne Overmyer

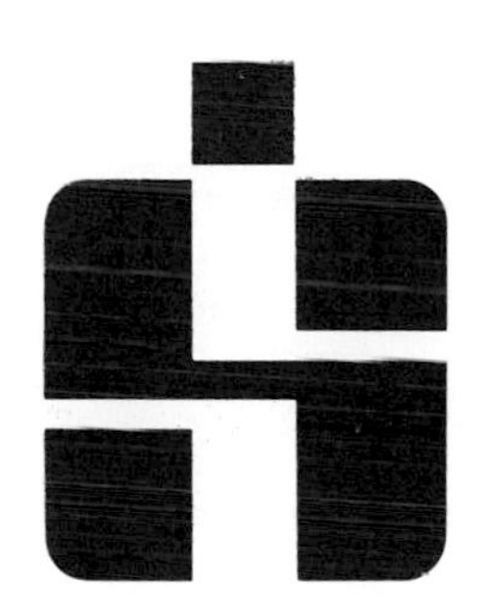

398

399

400

401

402

403

404	**Conceptual Systems Company**	Cleveland, Ohio *Electronic claim transmission system* Design: Ron Ray
405	**Cubicomp Corporation**	Berkeley, California *Computer graphics systems*
406	**Universal Data Systems**	Huntsville, Alabama *Modems*
407	**EEV Incorporated**	Elmsford, New York *Electron tubes*
408	**DisCopyLabs**	Santa Clara, California *Software duplication and downloading* Design: Evelyn Lim
409	**LOCUS**	Boalsburg, Pennsylvania *Electronics research & development* Design: Charles Bingham

404

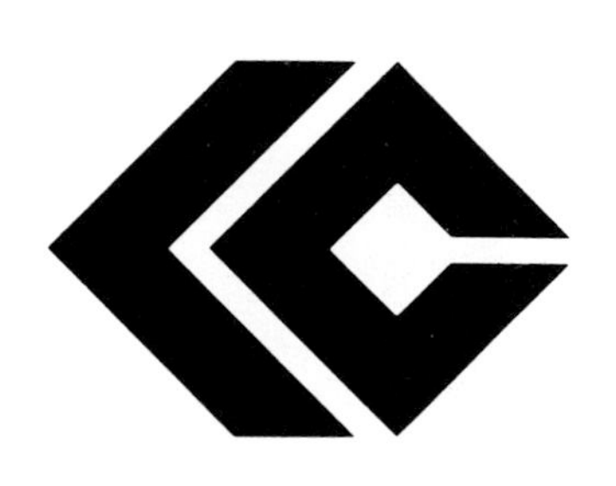

405

406

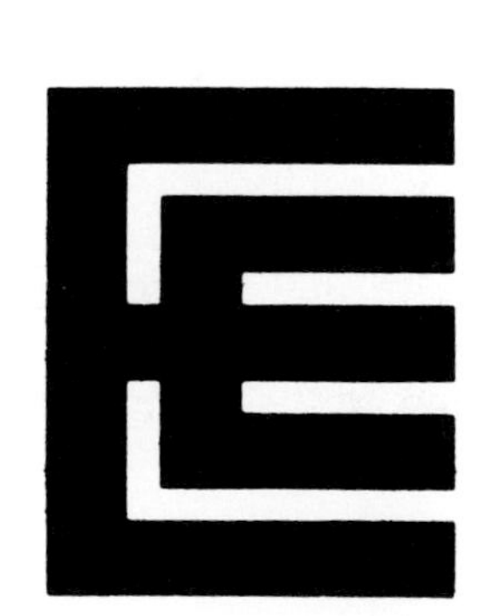

407

408

409

410	**Tritek Vision Systems**	Seattle, Washington *CAD software development* Design: Glenn Garner
411	**MicroPro International**	San Rafael, California *Business software*
412	**McHugh, Freeman & Associates**	Elm Grove, Wisconsin *Business & engineering software* Design: Don Itkin
413	**Ryan-McFarland Corporation**	Rolling Hills Estates, California *Computer software* Design: Reiser Williams DeYong
414	**International Technology Marketing**	Wellesley Hills, Massachusetts *CAD/CAM industry information service* Design: Roy Brown
415	**Monolithic Memories**	Santa Clara, California *LSI logic & memory systems* Design: In-house

410

411

412

413

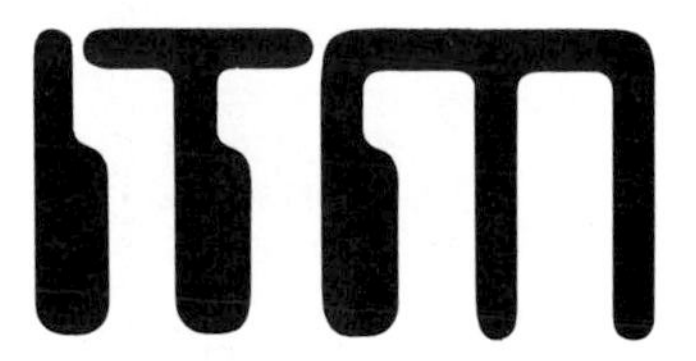

414

415

416 **QED Information Sciences**
Wellesley, Massachusetts
Computer education & PC tutorial software
Design: Carolyn P. Goodman

417 **Good Software Corporation**
Dallas, Texas
Finance software
Design: Apple Graphics

418 **Sloan Technology Corporation**
Santa Barbara, California
Deposition systems
Design: Gruyé Associates

419 **ASM Incorporated**
Palo Alto, California
Precision high technology fabrication
Design: Gruyé Associates

420 **Capital Equipment Corporation**
Burlington, Massachusetts
Computer hardware/software
Design: Kerry Newcom

421 **Veeco Instruments**
Plainview, New York
Semiconductors & industrial equipment

416

417

418

419

420

421

422	**TCS Software**	Houston, Texas *Business software* Design: In-house
423	**ERIC Software Publishing**	Fresno, California *Educational software* Design: Lone, Lord & Schon
424	**Phoenix Software**	Norwood, Massachusetts *Software engineering for microcomputers* Design: Nigberg Corporation
425	**T Maker Company**	Mountain View, California *Computer software* Design: Gassman-Mathis
426	**Groundstar Software**	Capitola, California *Microcomputer software*
427	**Actrix**	San Jose, California *Portable computers* Design: Jerry Blank

422

423

Phoenix

424

425

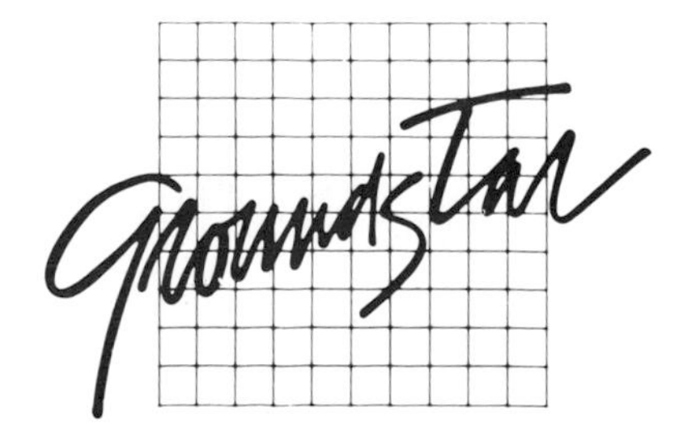

426

427

428	**Advanced Color Technology**	Chelmsford, Massachusetts *Jet printers* Design: Art 76
429	**Advantage Industries**	Orange County, California *Computerized property management services* Design: Susan Peterson
430	**Mu-Del Electronics**	Silver Spring, Maryland *RF and microwave communications equipment* Design: Heinz E. Blum
431	**Multisoft Corporation**	Beaverton, Oregon *Personal finance software* Design: In-house
432	**Develcon Electronics Ltd.**	Saskatoon, Canada *Electronic equipment* Design: Smail Communications
433	**Nittsuko Ltd.**	Tokyo, Japan *Telephone instruments* Design: Yusaku Kamekura

428

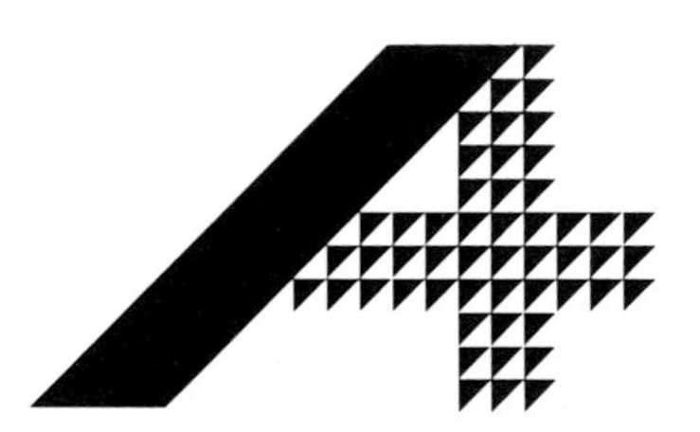

429

430

431

432

433

434 **The Mega Group**

Irvine, California
Software for IBM mainframes
Design: Lenae, Warford & Stone

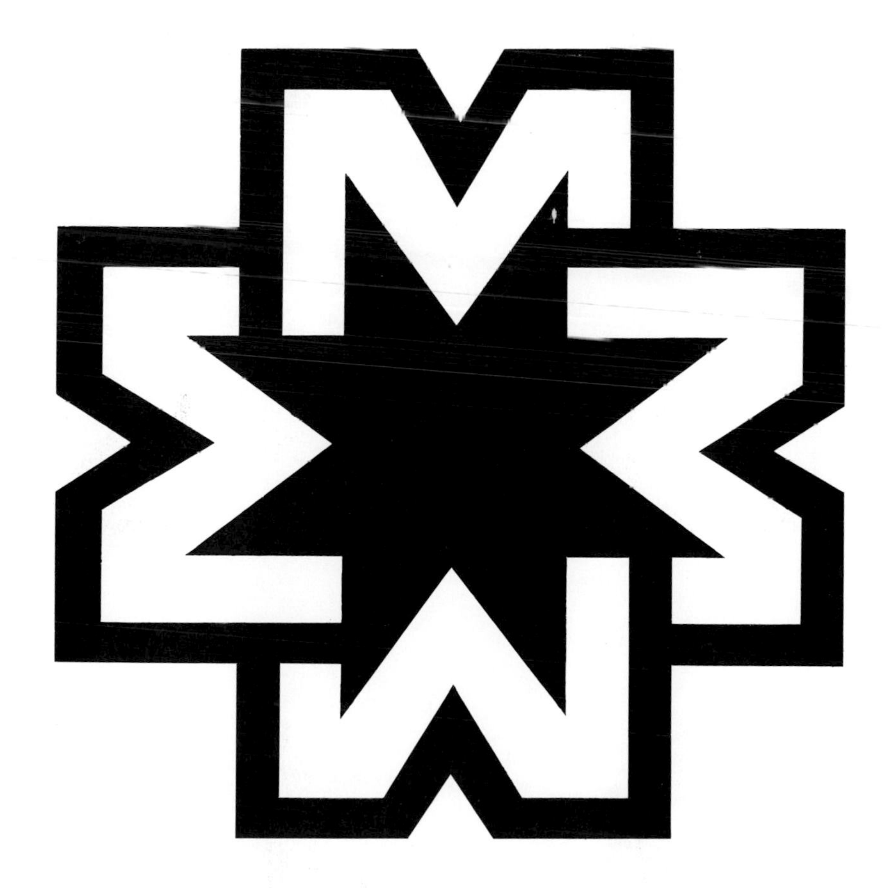

435 **GenRad Incorporated**
Waltham, Massachusetts
Automatic testing equipment
Design: Chermayeff & Geismar Associates

436 **GAI-Tronics Corporation**
Reading, Pennsylvania
Telecommunications equipment
Design: Alan M. Forrer

437 **GTCO, Corporation**
Rockville, Maryland
Electromagnetic digitizers
Design: Don Miller

438 **IQ Technologies**
Bellevue, Washington
Computer interface devices
Design: John Hegstrom

439 **Image Technology**
Littleton, Colorado
Board level products for the IBM PC
Design: David Price

440 **Linear Technology Corporation**
Milpitas, California
Linear integrated circuits

435

436

437

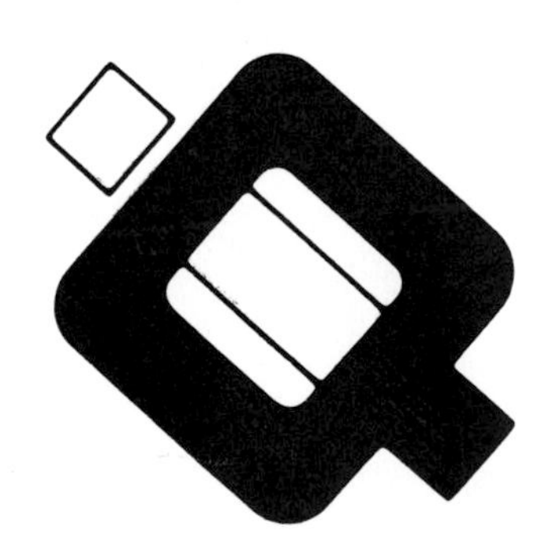

438

439

440

441	**CDR Systems**	San Diego, California *Harsh environmental instrumentation* Design: M. Brooks
442	**RanaSystems**	Chatsworth, California *Computer peripherals* Design: Huerta Design Associates
443	**Intermedia Systems**	Cupertino, California *Computer memory extension cards*
444	**Kinetic Systems**	Roslindale, Massachusetts *Air suspension & isolation systems* Design: Mike Sage
445	**Brookhaven National Laboratory**	Upton, New York *Physical & biomedical sciences research* Design: In-house
446	**Krohn-Hite Corporation**	Avon, Massachusetts *Electronic test and measurement instruments*

441

442

443

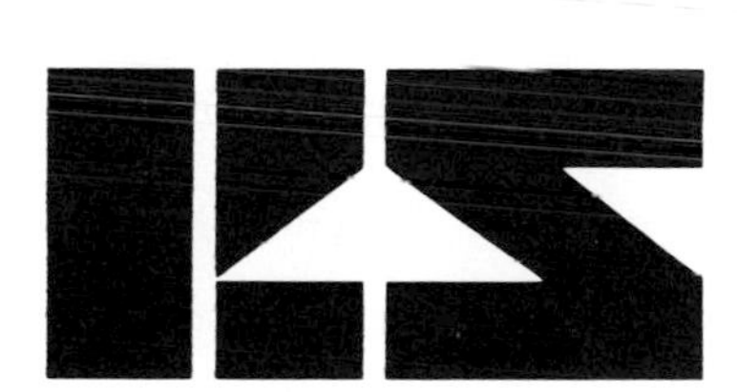

444

445

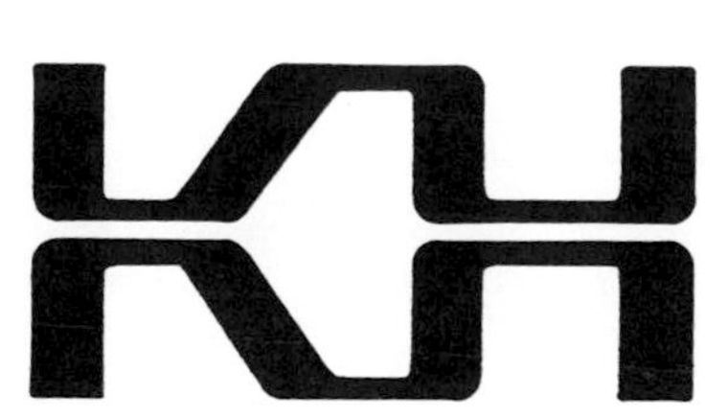

446

447	**Mad Computer**	Santa Clara, California *Personal computer workstations* Design: Mark Anderson Design
448	**ZAX Corporation**	Irvine, California *Microprocessor development systems*
449	**Oliver Advanced Engineering**	Glendale, California *Programmable semiconductor device testers* Design: Mark Wood
450	**Relational Database Systems**	Palo Alto, California *Database software* Design: Kevin Archer
451	**Cybervision**	Buffalo, New York *Graphics software packages* Design: Boris Stout
452	**Artificial Intelligence Research Group**	Los Angeles, California *Computer software*

447

448

449

450

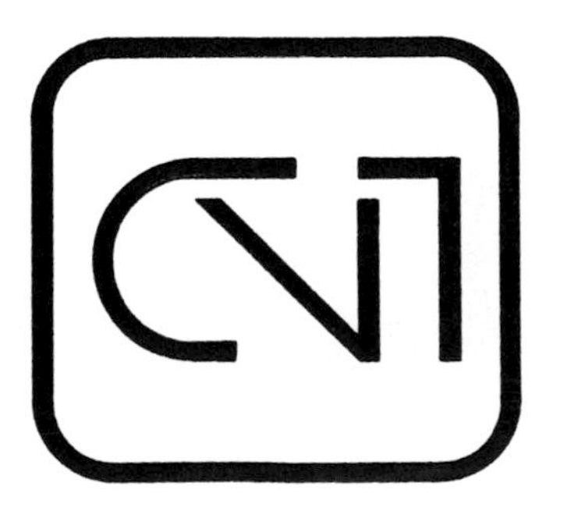

451

452

453	**Human Systems Dynamics**	Northridge, California *Statistical analysis software* Design: Media X
454	**DSD Corporation**	Kirkland, Washington *Computer software* Design: Floathe & Associates
455	**Scientific Marketing Services**	Minotola, New Jersey *Marketing services for technology companies* Design: Scientific Marketing Services
456	**AST Research**	Irvine, California *Enhancements for personal computers*
457	**Computers International Incorporated**	Los Angeles, California *Computer printers and word processors* Design: ME Graphics
458	**California Devices**	San Jose, California *Semicustom integrated circuits*

453

454

455

456

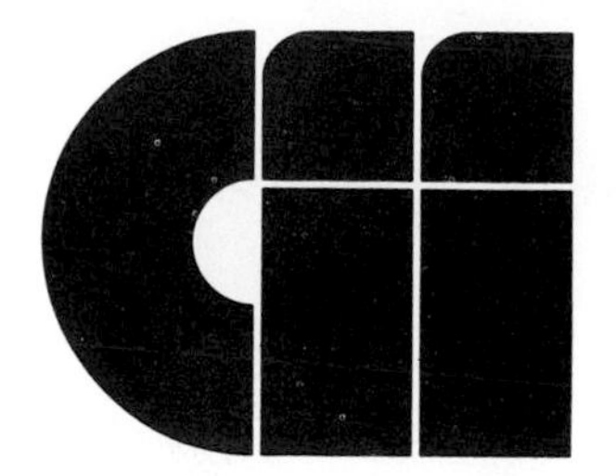

457

458

459 **Gail Epstein Enterprises**

Los Angeles, California
High tech contract furnishings
Design: Jerry L. Kuyper

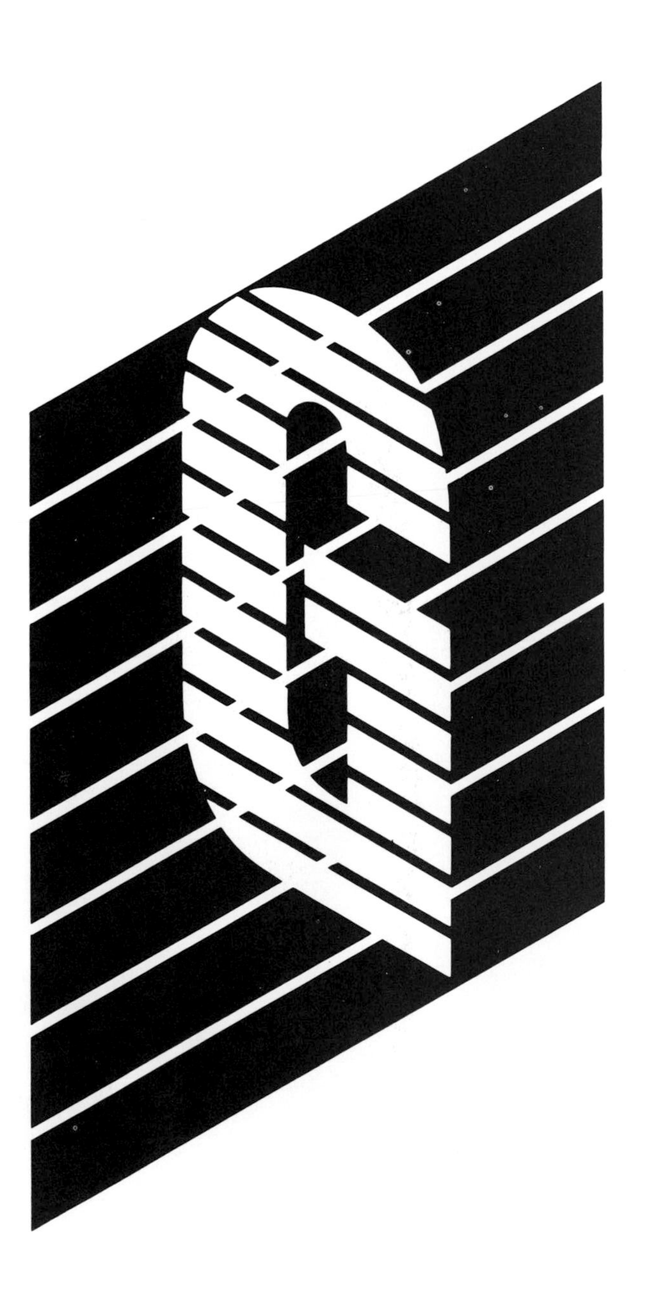

459

460	**Contec Sound Media Ltd.**	Hong Kong *Audio products* Design: Henry Steiner
461	**Innovative Data Technology**	San Diego, California *Computer tape drives & subsystems* Design: Reiser Williams & Long
462	**Applied Circuit Technology**	Anaheim, California *Computer disk and tape drive test equipment* Design: In-house
463	**RAF Software**	San Antonio, Texas *Computer software* Design: Jeremy Dack
464	**Unisource Software Corporation**	Cambridge, Massachusetts *Computer software* Design: The Niberg Corporation
465	**MLI Microsystems**	Framingham, Massachusetts *Computer software*

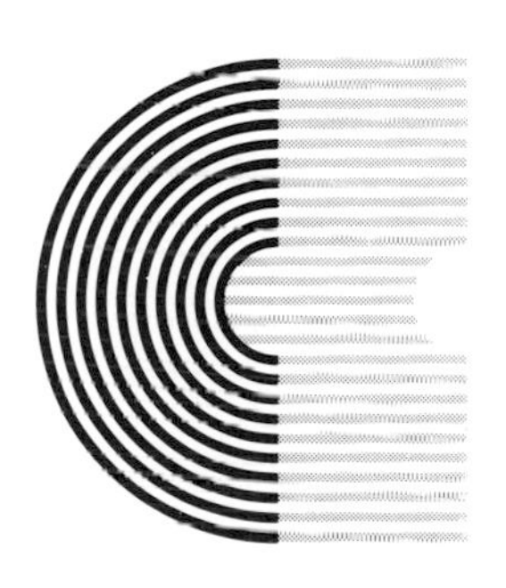

460

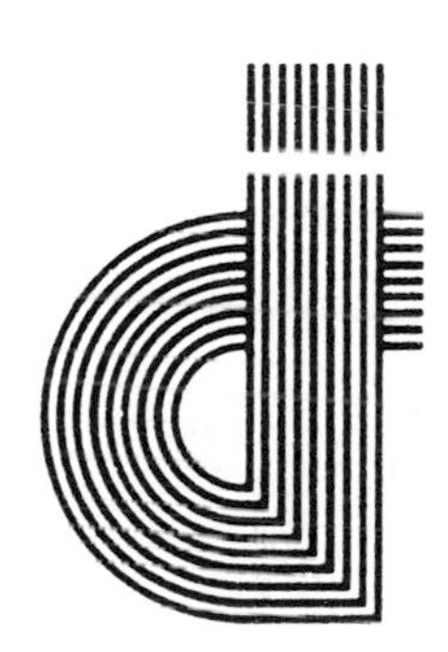

461

462

463

464

465

466	**P Cubed**	Wichita, Kansas *Software for the Apple Macintosh* Design: Kathy Lambert & Associates
467	**Microplot Systems**	Columbus, Ohio *Graphics software* Design: Stephen F. Bean
468	**Omnibyte Corporation**	West Chicago, Illinois *Single board computers* Design: Jon R. Burton
469	**Kierulff Electronics**	Tustin, California *Distributor of electronic components*
470	**Nanodata Computer Corporation**	Buffalo, New York *Mainframe digital computers* Design: In-house
471	**Penta Systems International**	Baltimore, Maryland *Automated text processing systems*

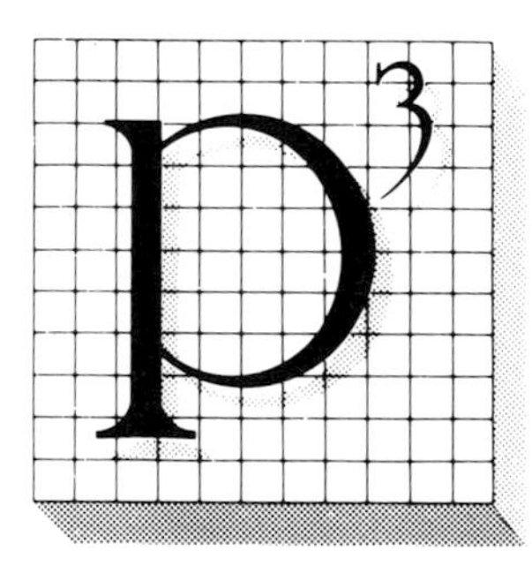

466

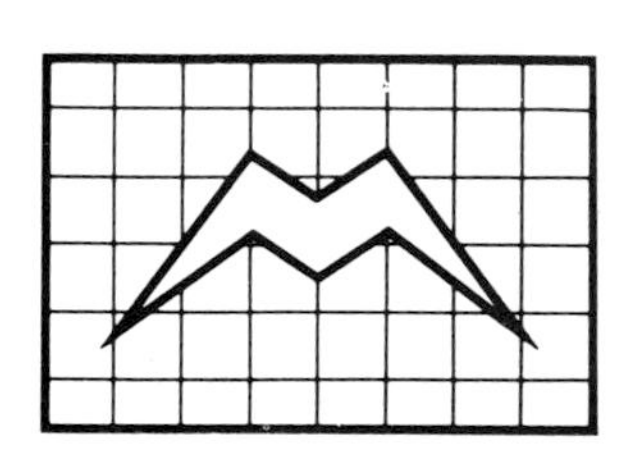

467

468

469

470

471

472	**Sigma Information Systems**	Anaheim, California *Disk controllers*
473	**Integral Quality Incorporated**	Seattle, Washington *Artificial intelligence development tools* Design: Robert F. Rorschach
474	**Computer Innovations**	Tinton Falls, New Jersey *Systems software*
475	**Heraeus Germalloy**	W. Conshohocken, Pennsylvania *Thick film & capacitor materials*
476	**Burndy Corporation**	Norwalk, Connecticut *Electronic & electric connectors* Design: Van Dyke Associates
477	**Elmwood Sensors**	Pawtucket, Rhode Island *Thermostats and heat sensor devices*

472

473

474

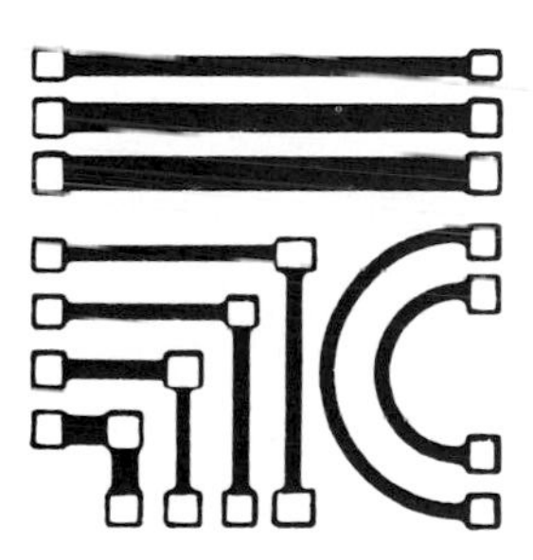

475

476

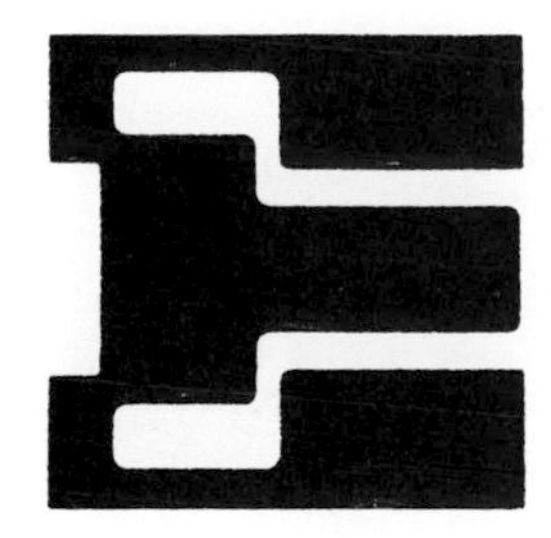

477

478	**Micro Flash Computer Systems**	Concord, California *Microcomputer sales* Design: Lekas Miller Design
479	**OSM Computer Corporation**	Mountain View, California *IBM compatible personal computers*
480	**Computer Accessories Corporation**	San Diego, California *Computer interface adaptor cables* Design: G'Marc Graphic Design
481	**Systemhouse Ltd.**	Ottowa, Canada *Computer software development* Design: Doug Seaborn
482	**Plasma-Therm Incorporated**	Kresson, New Jersey *Plasma process systems* Design: Scientific Marketing Services
483	**Veeder-Root Company**	Hartford, Connecticut *Industrial and fuel dispensing controls*

478

479

480

481

482

483

Aha Incorporated

Santa Cruz, California
Project management software
Design: Ron B. Swensen

485	**TDI Systems**	Rockville, Maryland *Wordstar® enhancement program* Design: Frank Ota
486	**Micro Data Base Systems**	Lafayette, Indiana *Software development for microcomputers* Design: Arrowhead Advertising
487	**BIT Software**	Milpitas, California *Computer software*
488	**Universal Industries**	Los Angeles, California *Computer furniture* Design: Studio 6
489	**Harris Microwave Semiconductor**	Milpitas, California *Gallium arsenide and microwave products*
490	**Digital Technology**	Dayton, Ohio *Computer support equipment*

485

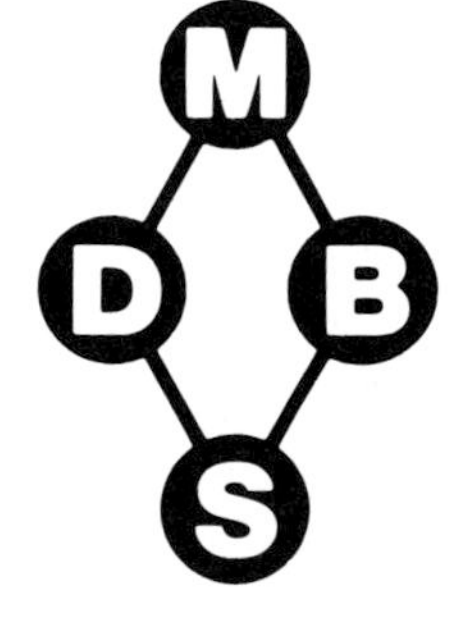

486

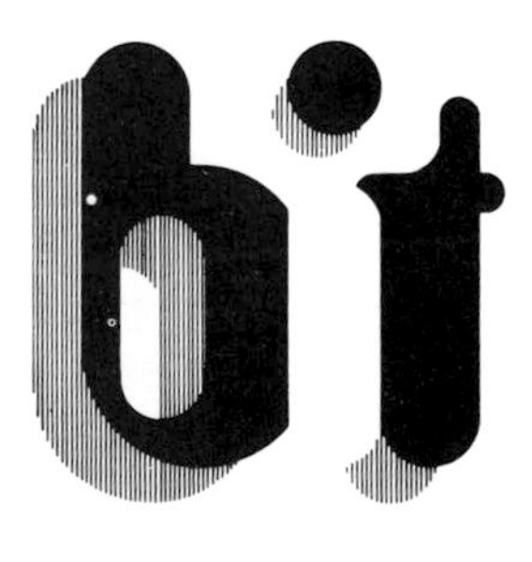

487

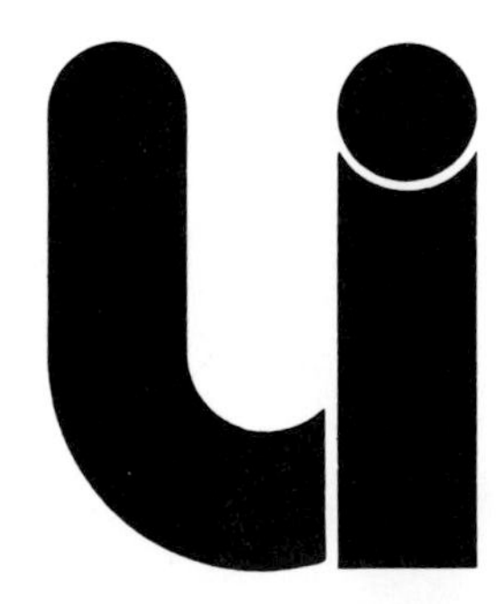

488

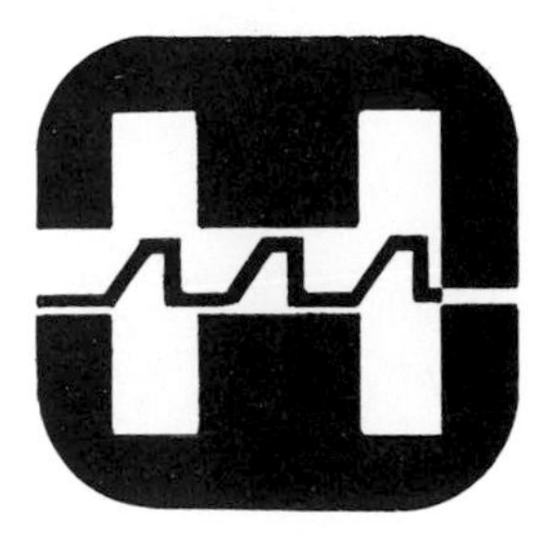

489

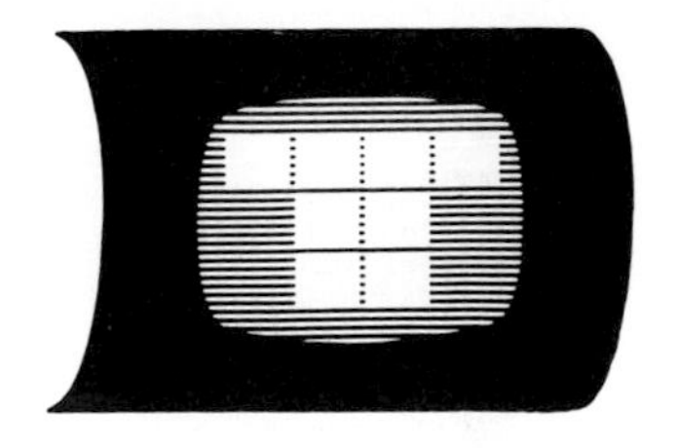

490

491	**Illumination Industries Incorporated**	Sunnyvale, California *Lighting systems* Design: Primo Angeli Graphics
492	**Mannesmann Talley Corporation**	Kent, Washington *Dot matrix printers*
493	**Execucom Systems Corporation**	Austin, Texas *Computer software* Design: Richards Brock Miller Mitchell & Associates
494	**Electric Transit**	Thousand Oaks, California *Computer games & interactive simulations* Design: COY
495	**Hercules Computer Technology**	Berkeley, California *Color graphics card for the IBM PC* Design: Kevin Jenkins
496	**Gelman Sciences**	Ann Arbor, Michigan *Membrane filtration products*

491

492

493

494

495

496

497	**Satellite Software International**	Orem, Utah *Software publisher* Design: Bob Simmons
498	**Plenary Systems**	Dallas, Texas *Accounting software* Design: BJW Marketing Communications
499	**Digital Communications Associates**	Norcross, Georgia *Data communications networking products* Design: Cole Henderson Drake
500	**Computer Software Design**	Anaheim, California *Computer software* Design: Chuck Bennett
501	**TCI Software**	Flourtown, Pennsylvania *Computer software*
502	**Tseng Laboratories**	Newton, Pennsylvania *Integrated enhancement add-on boards*

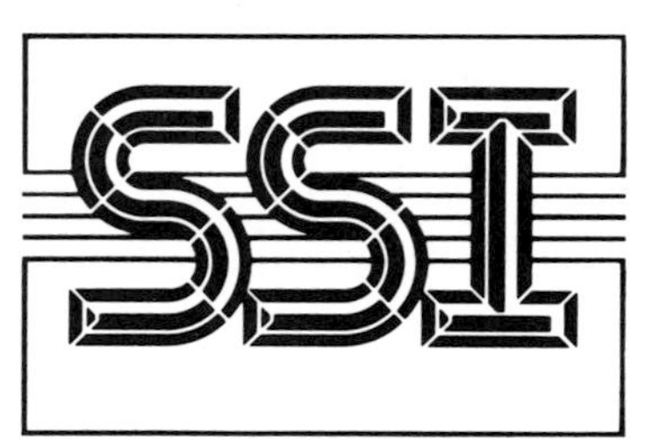

497

498

499

500

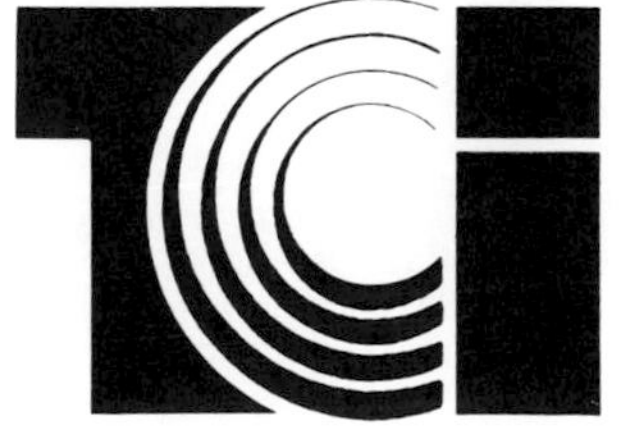

501

502

503	**Recruit Computer Print**	Tokyo, Japan *Computer division of Recruit Company* Design: Yusaku Kamekura
504	**NEC Corporation**	Hawthorne, California *Electronic & communications equipment*
505	**LEJ Incorporated**	Paramus, New Jersey *Microcomputer peripherals and accessories* Design: Tony Barnes
506	**Floating Point Systems**	Beaverton, Oregon *Array processors & scientific computers* Design: Rudi Milpacher
507	**Advanced Processor Systems**	Austin, Texas *Microcomputer peripherals* Design: Curtis P. Elmore
508	**Instruments SA**	Metuchen, New Jersey *Electronic devices*

503

504

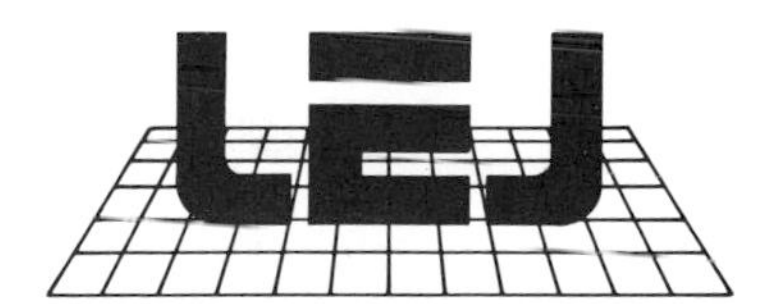

505

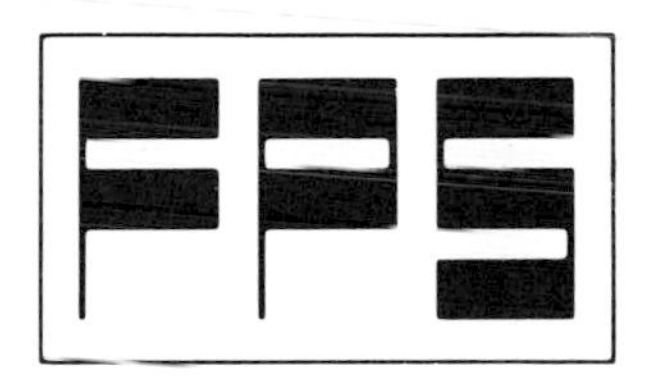

506

507

508

509 **Ability**

Toronto, Ontario, Canada
Integrated software system
Design: Spencer/Francey Incorporated

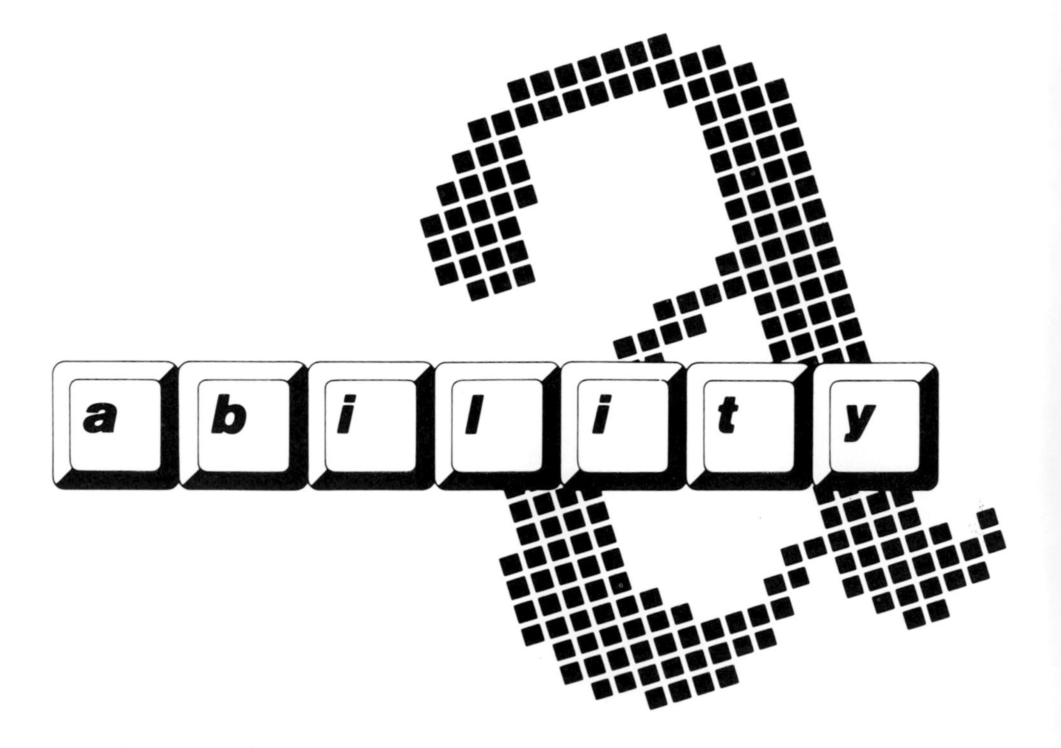

510	**Dylon Data Corporation**	San Diego, California *Magnetic tape controllers* Design: Jon Scharnborg & Associates
511	**Data Systems for Industry**	Carson, California *Manufacturing management software* Design: Don Whipple
512	**Xytel**	Lombard, Illinois *Microcomputer controls* Design: Jan Lorenc Design
513	**Chips and Technologies**	San Jose, California *Integrated circuits* Design: Lawrence Bender & Associates
514	**Logical Devices**	Ft. Lauderdale, Florida *Software drivers*
515	**Digital**	West Concord, Massachusetts *Computer hardware/software*

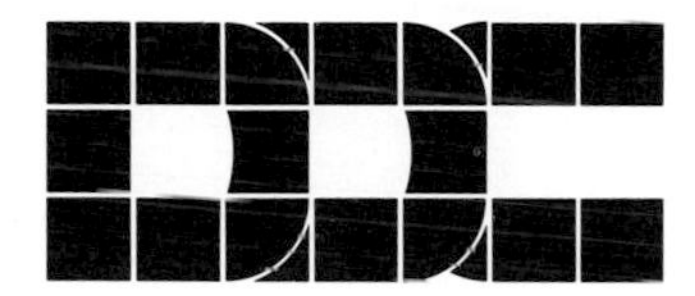

510

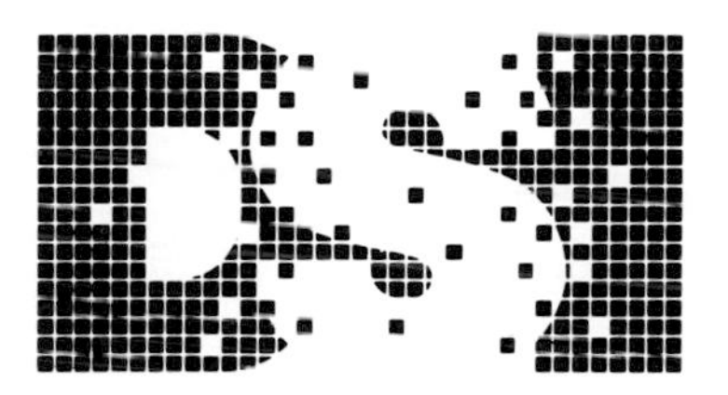

511

512

513

514

515

516 **Aktis product logo**

Sunnyvale, California
Medical laboratory test equipment
Design: Jerry Blank

517 **Spectra Logic**

Santa Clara, California
Controllers
Design: Alice Baher McGown

518 **Emerging Technology Consulting**

Boulder, Colorado
Software development
Design: Jennifer Pollman

516

517

518

519 **Atkis**
Sunnyvale, California
Test equipment for the medical laboratory
Design: Jerry Blank

520 **Alloy Computer Products**
Framingham, Massachusetts
Tape subsystems for personal computers

521 **Micro Education Corporation of America**
Westport, Connecticut
Computer software
Design: Hank W. McIver

522 **Intel Corporation**
Santa Clara, California
Microprocessors & semiconductors

523 **Geri Engineering**
San Mateo, California
Technical research & development
Design: Gruyé Associates

524 **Grid Systems Corporation**
Mountain View, California
Portable computers
Design: Primo Angeli Graphics

519

520

521

522

523

524

525	**Plexus Computers**	San Jose, California *UNIX based minicomputers* Design: Lawrence Bender & Associates
526	**FORTIS / Dynax Incorporated**	Bell, California *Color monitors* Design: Shinya Kano
527	**Cascade Graphics Development**	Santa Ana, California *Computer-aided drafting turnkey systems* Design: Ripley-Woodbury

PLEXUS

525

526

CASCADE

527

528	**Parallax Graphics**	Sunnyvale, California *Controller boards for computer graphics* Design: Jerry Blank
529	**Mall Telecommunications**	Mountain View, California *Business telephone systems* Design: Lawrence Bender & Associates
530	**TASA**	Santa Clara, California *Touch activated switch arrays* Design: Gruyé Associates

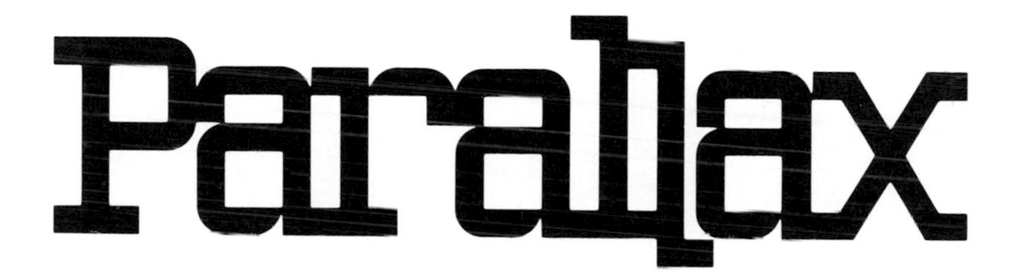

528

529

530

531	**Windfarms**	Los Angeles, California *Wind energy development* Design: April Greiman
532	**Tektronix**	Santa Clara, California *Electronic test & measurement equipment* Design: S&O Consultants
533	**Raytheon**	Lexington, Massachusetts *Electronic equipment* Design: Anspach Grossman Portugal

WINDFARMS

531

Tektronix

532

Raytheon

533

534	**Adax Corporation**	Atlanta, Georgia *Computer/telephone interfaces* Design: Beth Tippett
535	**Mindset Corporation**	Sunnyvale, California *Compyter hardware/software* Design: Michael Vanderbyl
536	**Userview / Product logo**	Atlanta, Georgia *Data management software* Design: Toby Hufham
537	**Inmos Ltd.**	Bristol, England *Integrated circuits & computer software* Design: Don Burston & Associates
538	**Juki Industries of America**	New York, New York *Computer printers* Design: The Alden Group
539	**Okidata**	Mount Laurel, New Jersey *Dot matrix printers*

534

535

536

537

538

OKIDATA

539

540	**Inotek**	Dallas, Texas *Computer hardware sales* Design: Richards Brock Miller Mitchell & Associates
541	**NeoLogik**	Los Gatos, California *Voice response systems for computers* Design: In-house
542	**Phaser Systems**	San Francisco, California *Productivity enhancement software* Design: Pickett Communications
543	**John Fluke Manufacturing Company**	Everett, Washington *Electronic testing and measuring instruments*
544	**Entrepo**	Sunnyvale, California *Microdrives and tapes* Design: Carole Taylor
545	**Compaq Computer Corporation**	Houston, Texas *Personal computers*

540

541

542

543

544

545

546	**Xanaro Technologies**	Toronto, Ontario, Canada *Integrated software* Design: Spencer/Francey Incorporated
547	**Corvus Systems**	San Jose, California *Computer workstations & peripherals* Design: S & O Consultants
548	**PC Ware**	San Jose, California *Hardware peripherals for personal computers* Design: Blakeley Graphics
549	**Wyse Technology**	San Jose, California *Microprocessor workstations* Design: Mark Anderson Design
550	**KnowledgeMan / Micro Data Base Systems**	Lafayette, Indiana *Knowledge management system software* Design: Guldberg & Cooper Agency
551	**MOM Corporation**	Atlanta, Georgia *Micro-to-mainframe interface systems* Design: Weir & Pfeiffer

XANARO

546

547

548

549

550

551

552 **Cambridge Digital**
Cambridge, Massachusetts
Computer system integration
Design: The Nigberg Corporation

553 **Inovision**
Dallas, Taxes
Electronic data systems
Design: Richards Brock Miller Mitchell & Associates

554 **Springboard Software**
Minneapolis, Minnesota
Educational software
Design: Madsen & Kuester

Cambridge
Digital

552

553

554

555 **Persoft**
Madison, Wisconsin
Terminal evaluation/file transfer software
Design: Waldbillig & Bestman

556 **Integra Signum AG**
Wallisellen, Switzerland
Security systems
Design: Josef Müller-Brockman

557 **Microstuf / Product logo**
Atlanta, Georgia
Data communications software
Design: Toby Hufham

555

integra signum

556

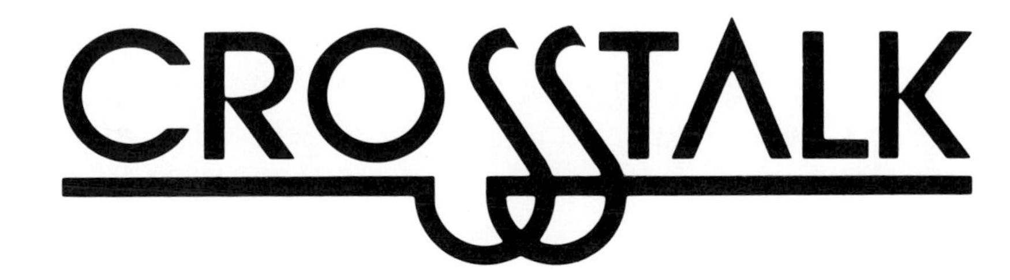

557

558	**Vectrix**	Greensboro, North Carolina *High resolution computer terminals* Design: In-house
559	**Index Technology Corporation**	Cambridge, Massachusetts *Systems analysis & design software* Design: Gill Fishman Associates
560	**Best Programs Incorporated**	Alexandria, Virginia *Computer software* Design: In-house
561	**IMSI**	San Rafael, California *Business software* Design: Tony Naganuma
562	**IQ Computer Corporation**	Fort Worth, Texas *Offshore management of computer products* Design: Roberson Illustration
563	**Domus Software Ltd.**	Ottawa, Ontario, Canada *Software*

558

559

560

561

562

563

Appendix

Menu icons for the Apple Macintosh personal computer

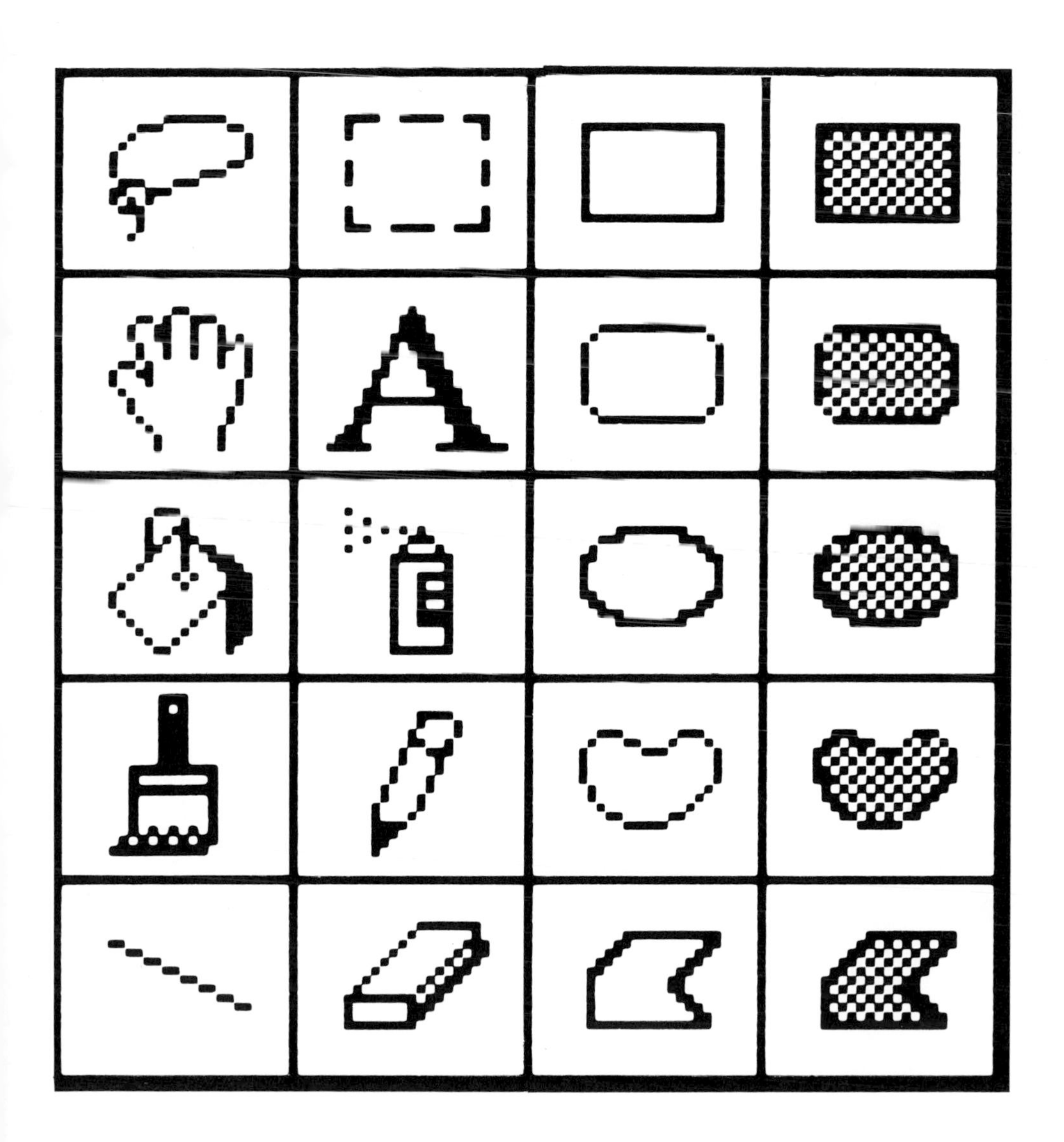

Symbols for IBM dictating equipment controls designed by Paul Rand

record

output

foot pedal

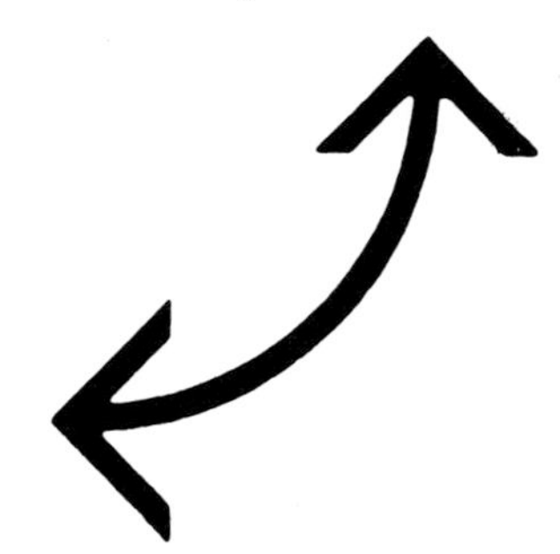

off

speaker

auxiliary power

volume

conference

instructions to sec'y

listen

mic-listen

microphone

reverse

dictate

tone

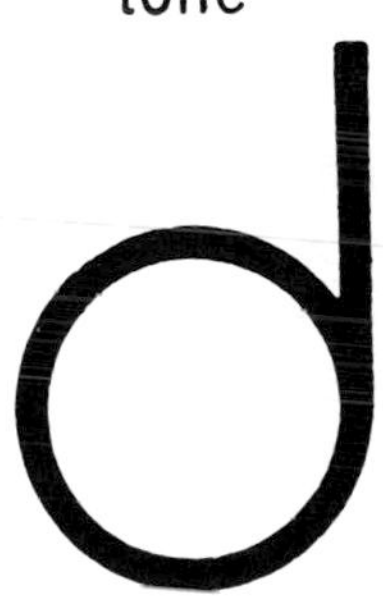

tuning

foot switch

end of letter

Video wipe patterns utilized in computer-assisted graphics animation. One image dissolves to the next with the use of a video switcher. ***(Grass Valley 1600 Series)***

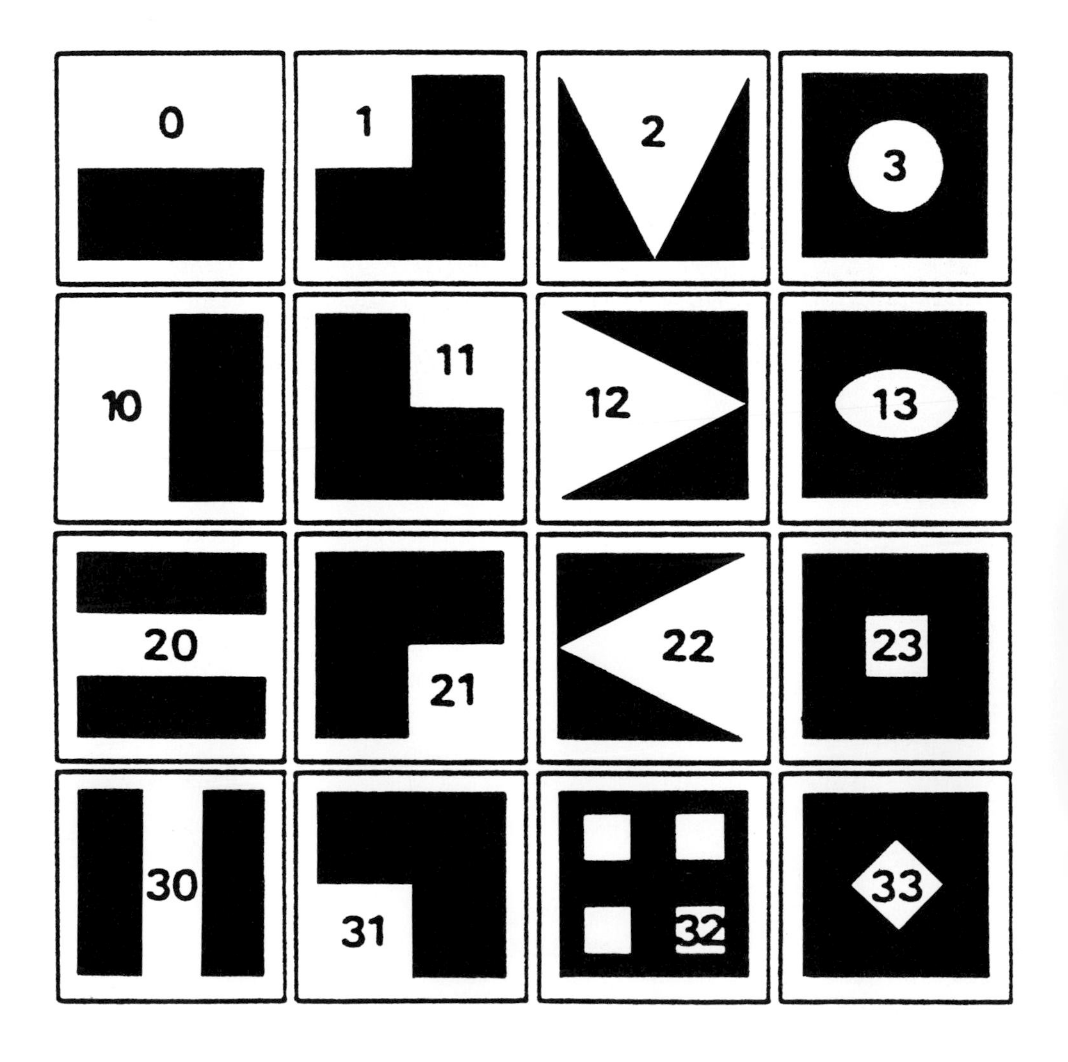

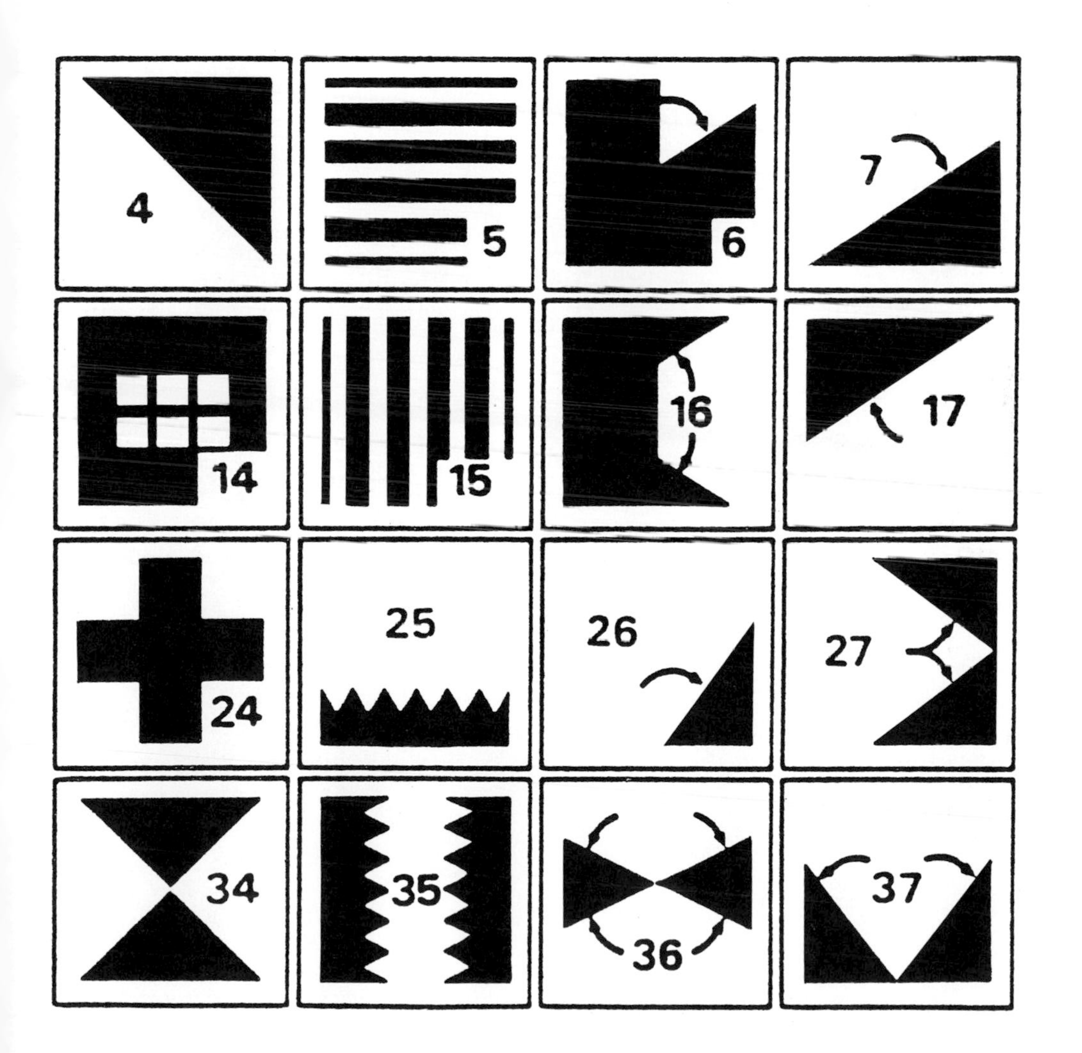
4
5
6
7
14
15
16
17
24
25
26
27
34
35
36
37

A 63 character set of 7 X 7 dot matrix characters
(Centronics Data Computer)

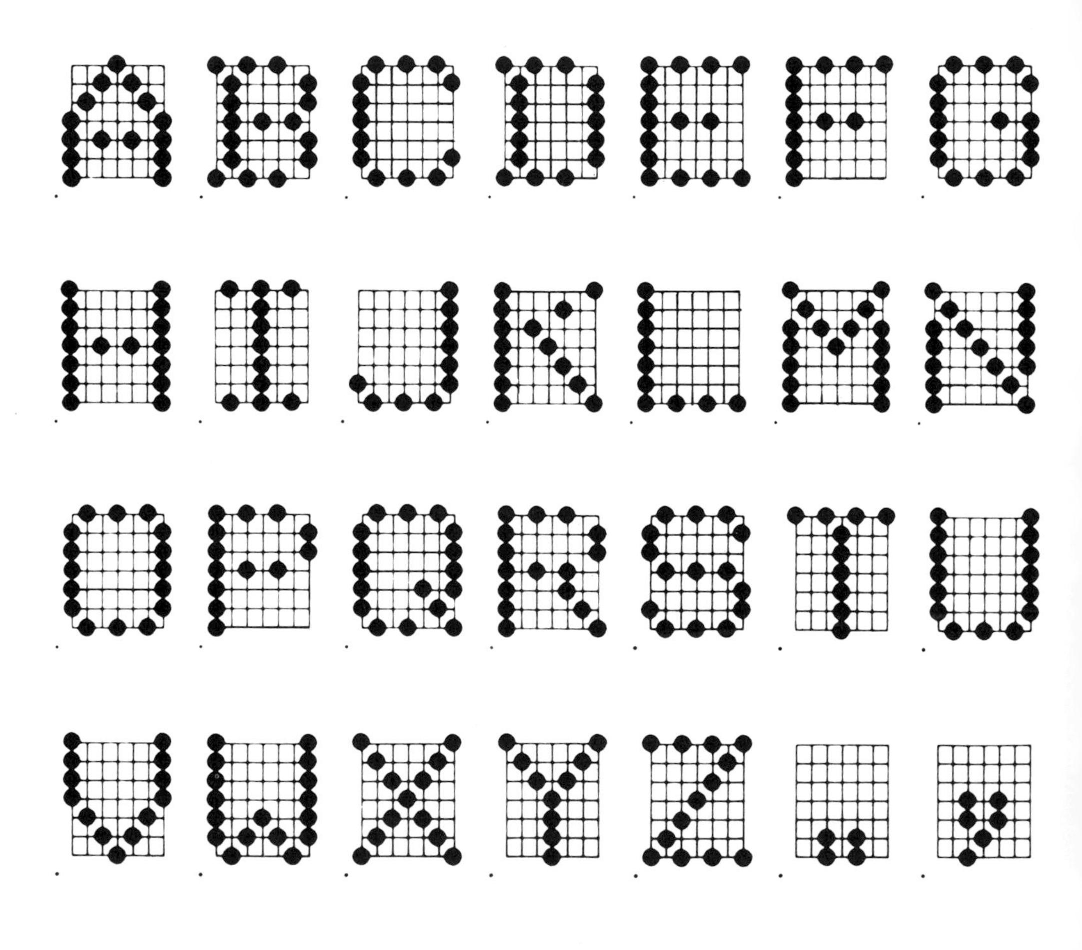

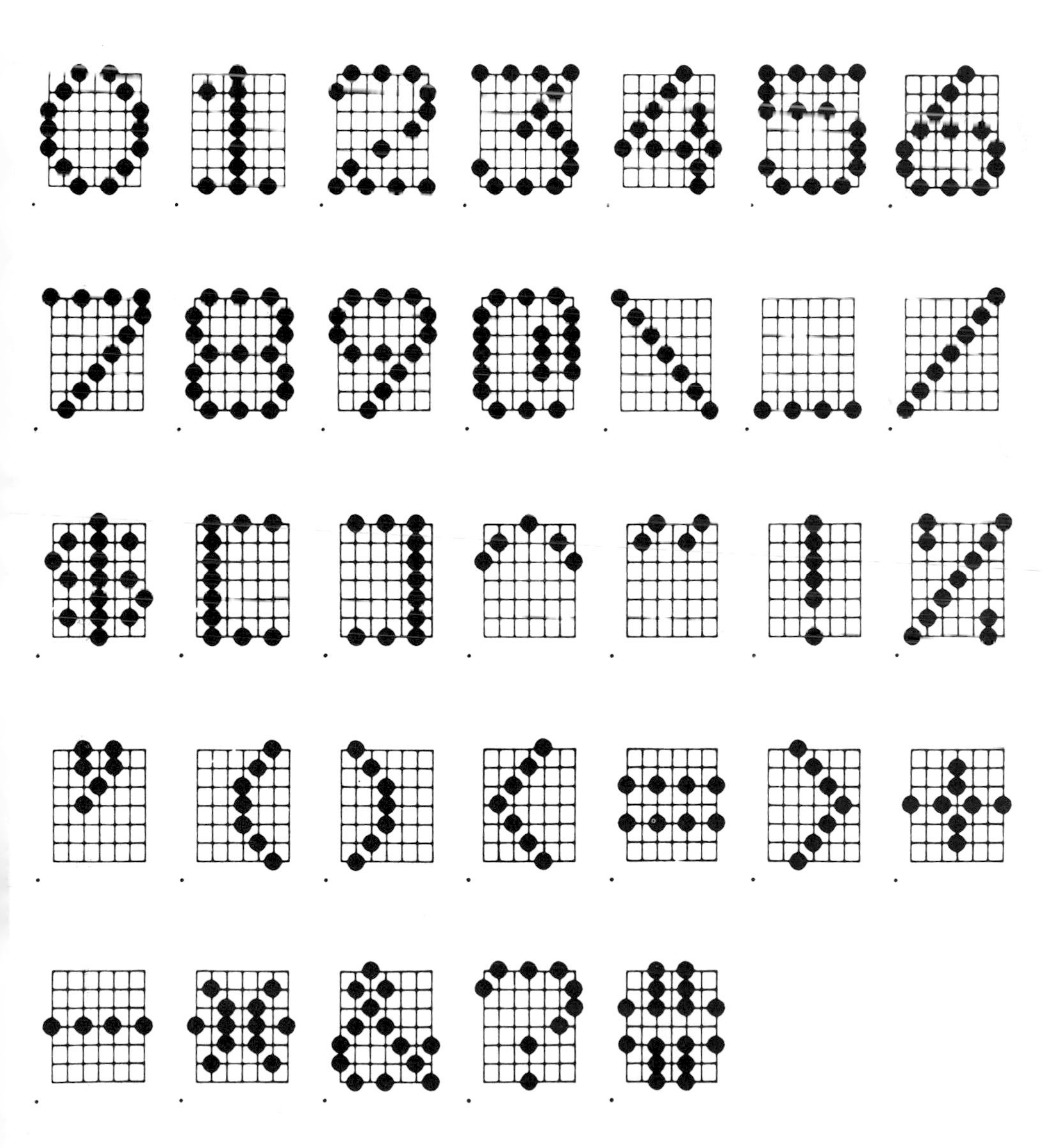

Bibliography

American Trademarks 1930-1950.
John Mendenhall.
Art Direction: New York, 1983.

Dictionary of Symbols.
J.E. Cirlot.
Philosophical Library: New York, 1962.

Design & Drafting of Printed Circuits.
Darryl Lindsey.
Bishop Graphics: Westlake Village, 1982.

MacPaint Manual.
Apple Computer; Cupertino, 1983.

McGraw-Hill Computer Handbook.
Harry Helms, Editor.
McGraw-Hill: New York, 1983.

Pratt Center for Computer Graphics In Design.
Seminar: New York, 1983.

Index to Companies

Index to Designers